THE COCKROACH COMBAT MANUAL

Quill

New York

Designed by Robert Powell

Library of Congress Cataloging in Publication Data

Frishman, Austin M
  The cockroach combat manual.

  "Morrow quill paperbacks."

  Includes index.
  1. Cockroaches—Control. I. Schwartz, Arthur P., joint author. II. Title.
[TX325.F75 1980b]   648'.7   80-288
ISBN 0-688-03613-9
ISBN 0-688-08613-6 pbk.

Printed in the United States of America

2 3 4 5 6 7 8 9 10

THE LOATHSOME COCKROACH

i am going to start
a revolution
i saw a kitchen
worker killing
water bugs with poison
hunting pretty
little roaches
down to death
it set my blood to
boiling
i thought of all
the massacres and slaughter
of persecuted insects
at the hands of cruel humans
and i cried
aloud to heaven
and i knelt
on all six legs
and vowed a vow
of vengeance*

*From *The Lives and Times of Archy and Mehitabel*, by Don Marquis.

CONTENTS

Chapters 1, 2, 5, and 6 by permission of Austin M. Frishman and the University of the State of New York, Stony Brook, Long Island. Written by Austin M. Frishman, Jan Gerschkoff, Lone Kelstrup and Arthur Schwartz.

Charts in Chapter 9 by permission of the Whitmire Research Laboratories, Inc., St. Louis, Missouri.

Crawling through cockroach history

1. THE EVOLUTION OF THE COCKROACH

Three hundred fifty million years ago cockroaches thrived. This age was called the Carboniferous Period. Scientists refer to it as the Age of Cockroaches because at that time these insects had reached their peak in number of species as well as abundance of each species. The presence of a warm, moist environment allowed these cockroaches to dominate; today none of these species exist. However, the current-day species that evolved from this now-extinct group are called Paleoblattidae. *Blattidae* is the Latin term used to describe present-day cockroaches, and the word *Paleo* refers to the term ancient. The Carboniferous Period came toward the end of the Paleozoic Era, which began 500 million years ago and ended 200 million years ago. Dinosaurs had not begun to roam the earth.

Some experts recognize twelve families of fossil cockroaches. The largest family is called the Archimylacridae and comprised as many as 350 species.

Surprisingly, fossil remains indicate that there have been very few structural changes. The shape of the cockroaches' bodies and their fondness for living in moist places have changed little. Domestic cockroaches that exist today have evolved to the point where they can withstand drier conditions. Their wings have also become reduced in size, and their eggs are now deposited in a capsule to reduce dehydration. Primitive cockroaches would deposit their eggs singly in a moist environment.

Fossil of cockroach

Coming to America via slave ship

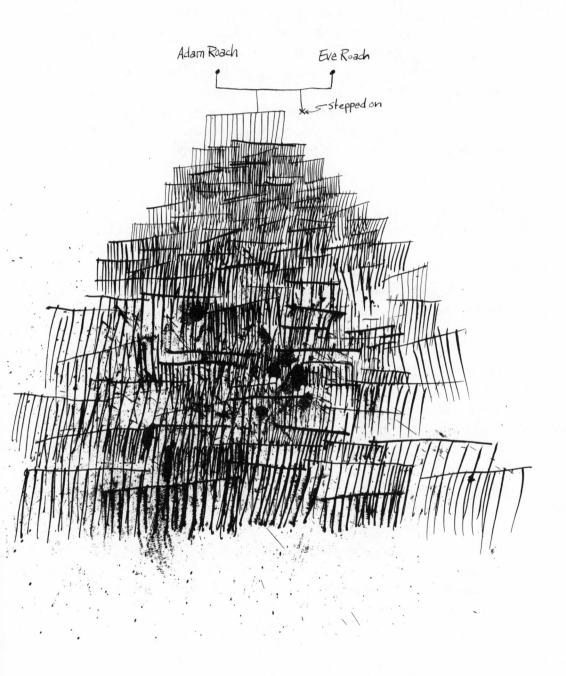

The cockroach family tree

The sword-shaped ovipositor evolved over a period of time to be retracted into the abdomen of the male cockroach and into the formation of a genital pouch in the female. In the primitive period, the function of the ovipositor was to inject eggs into the soil or other suitable mediums. Today, the ovipositor is used to guide eggs into an ootheca. An ootheca is an egg capsule which houses the eggs and is produced by the female. Of the domestic cockroaches, with the exception of the German cockroach, the egg capsule is dropped or glued to surfaces. This waterproof structure protects the eggs and allows these creatures to encroach upon man.

There are two main locations in North America where scientists collect fossil cockroaches: the Middle or Lower Measures of Illinois, and the Upper Coal Measures in Kansas.

The dominant cockroach today is the German cockroach or *Blattella germanica* (Linnaeus). Originally this cockroach evolved in North Africa. It is believed to have migrated to eastern Europe in Greek and Phoenician ships, whereby it then spread to Asia Minor, the Black Sea region and southern Russia. According to P. B. Cornwell, a British entomologist and director of research for Rentokil Laboratories Ltd., East Gunstead, Sussex, England, this species then spread northward and westward across Europe. Once entrenched in western Europe, it was just a matter of time before the German cockroach successfully invaded ships. Today these world travelers go first-class via planes, ships, trains and buses. The three main factors which contribute to the dominance of the German roach are its small size, short life cycle and prolific breeding capabilities.

Over the centuries, individual roaches within a given species mutate. Environmental conditions determine which mutants are selected to survive. Cockroaches have exhibited an extraordinary record of successes, including the ability to develop a resistance to pesticides, withstand high dosages of radiation and survive on a minimal food intake.

Cockroaches Through History

The cockroach has been known to man throughout history. Household cockroaches are as ancient as the concept of food in the home. Those of us who have these pests are simply following time-honored traditions and precedent. So, if you have them don't be too upset.

It is believed that the four major household species of cockroaches all came from Africa. Some made the roundabout trip through the Mideast to Europe in trade caravans, or across the Mediterranean in Greek and Phoenician ships, arriving centuries later in North America. Others traveled directly across the Atlantic in slave ships. This latter group includes the so-called American cockroach.

References to the cockroach prevail through history. Dioscorides Pedanius, a Greek who was army physician for the Emperor Nero, stated in the first century that cockroach entrails, when mixed with oil and stuffed into the ear, would cure earaches. This information was included in his treatise on medicine called *De Materia Medica*. For sixteen centuries, *De Materia Medica* was considered the highest authority on medicine and was universally studied by medical students and botanists.

The cockroach is also mentioned in another first-century manuscript, a scientific encyclopedia called *Natural History*. Written by Pliny the Elder, it contained almost the same recipe for earaches as that of Dioscorides, but slightly improved: The oil that was mixed with the cockroaches had to be rose oil. To Pliny, a naturalist, cockroaches were a cure-all. When crushed, he observed, they would cure itching, tumors and scabs; when ingested with their wings and feet cut off, they relieved swollen glands.

As noted earlier, cockroaches were among the original travelers on sailing ships. Cockroaches at sea are described in the Danish Navy Annals of 1611 A.D. From time to time, apparently, cockroach hunts were held. The price for a thousand cockroaches was a bottle of brandy from the cook's locked pantry. Cockroach time was party time in those days. The Navy Annals record a single catch of this sort once bringing in 32,500 cockroaches.

The Spanish Armada had the same problems. Sir Francis Drake reported that upon capturing the *San Felipe*, his boarding

The cockroach's role in medicine

party found the bombarded decks overrun with cockroaches.

Cockroaches aboard ships have always manifested themselves in unpleasant ways. For instance, R. H. Lewis wrote that on a voyage from England to Tasmania:

Hundreds of cockroaches were flying around my cabin. They were in immense confusion and had communication with every part of the ship, between the timbers or the skin. The ravages they committed on everything edible were very extensive. Not a bisquit but was more or less polluted by them, and amongst the cargo of 300 cases of cheese, which had holes in them to prevent their sweating, were considerably damaged, some of them being half devoured—and not one without some marks of their residence.

British entomologist A. Nichols, among others, reported that sailors frequently complained of having their toenails, fingernails and the calloused parts of their hands and feet nibbled by cockroaches. Also reported was the consumption of eyelashes and hair from sleeping crews and passengers, especially children.

Caudell, another entomologist on an explorer's expedition to British Columbia, reported: "On this trip I had them served to me in three different styles. Alive in strawberries, a la carte with fried fish—and baked in bisquit."

As late as the sixteenth century, physicians were still prescribing cockroach entrails as an excellent remedy for sore ears. A Viennese physician by the name of Matthiole updated Dioscorides' recipe of 1,600 years earlier: The cockroach entrails had to be cooked in oil instead of just mixed. In 1725, Jamaican children were still being fed cockroaches as a worm cure. The fact is, reports of cockroaches being used medicinally and nutritionally persist around the world even today.

In the summer of 1979, over a million German cockroaches were found thriving in a two-family house in Schenectady, New York. T-shirts were quickly distributed with the words "Schenectady—Cockroach Capital of the World."

The cockroach has frequently been privileged to be a social issue. In Northern Germany it is referred to as "Schwabe," a term for inhabitants of Southern Germany. In the south the cockroach is popularly known as "Preusze," after the Northern Germans. This

The cockroach's role on the seven seas

species of cockroach is known in the United States even today among exterminators as the German cockroach. The names "Yankee Settler," "Croton Bug" and "Bombay Canary," to name a few, have also been used at one time or another.

Cockroach history would not be complete without mentioning its debut in the legal profession. Fifty years ago a renowned Danish baker was faced with a trial when a customer found a cockroach in a piece of pastry. The baker, visualizing the end of his career, asked in court to see the piece of evidence. It was handed to him and he exclaimed, "Cockroach? That is not a cockroach, that is a raisin . . ." and swallowed it. He was acquitted.

What is it about the cockroach that makes it objectionable to human beings? Is it simply an aesthetic inconvenience? We know that eradication attempts have been made throughout the ages. For instance, Captain William Bligh, in his 1792 chronicling of the voyages of the H.M.S. *Bounty*, described his attempts to rid the ship of cockroaches using boiling water.

W. S. Blatchley described the following common method used in Mexico: "To get rid of cockroaches—catch three and put them in a bottle, and so carry them to where two roads cross. Here hold the bottle upside down, and as they fall out repeat aloud three credos. Then all the cockroaches in the house from which these three came will go away."

And as late as 1905, the Japanese Navy was still using the seventeenth-century Danish method of eradication. They called it "Shore Leave for Cockroaches." A Japanese seaman needed to capture only 300 roaches to be granted a day's shore leave. With characteristic succinctness and aplomb, the Japanese described the purpose as follows: "To promote extermination of cockroaches in a warship because, on the one hand, any warship suffers from numerous cockroaches, and on the other hand, any seaman likes shore leave."

Cockroaches are objectionable for more than aesthetic reasons. Aside from their being unwelcome visitors of minute size, great numbers and unpredictable direction and speed of flight, cockroaches happen to be contaminators of food supplies. If you have never smelled their odor, you are lucky. Called "attar of roaches," it is the combined product of their excrement, of fluid exuded from

The cockroach's role in culinary delights

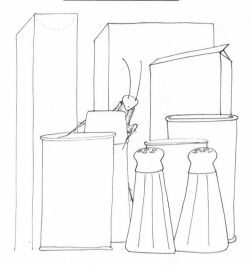

their abdominal scent glands and of a dark-colored fluid regurgitated from their mouths while feeding. This attar fouls food.

Cockroaches are a proven health hazard. The evidence was circumstantial for the longest time, but recent research in the United States has provided direct evidence that cockroaches are capable of carrying harmful microorganisms from unsanitary conditions to humans and their food supplies. It is now known that cockroaches may carry either inside or on the surface of their bodies pathogens for the following ailments:

Abscesses
Boils
Bubonic plague
Diarrhea
Dysentery
Food poisoning
Gastroenteritis
Intestinal infections
Leprosy
Lesions
Typhoid fever
Urinary tract infections

After countless ages we now know the cockroach to be more unwelcome in man's company than it has ever been before.

23

2. THE SOCIOLOGY OF THE COCKROACH

Cockroaches are uncomplicated creatures. They need little and ask for little. They are somewhat gregarious and may hide together in clusters. They share no outward affection with each other. Each will exist or coexist in what it finds to be the most comfortable environmental niche. Cockroaches are basically survival machines that hide by day and forage for food by night.

Cockroaches do not get thrills out of surprising or scaring people. They are simply trying to hide. That's why one may be discovered unexpectedly when something is moved. If you see one out of hiding or out in the open during the day, it probably means either it is hungry or there are too many other cockroaches hiding for it also to find a place to hide. That's the story on spotting cockroaches out of hiding.

If one can take a detached look at the cockroach, there are some interesting observations that can be made: As an insect, it holds a unique position in all the animal kingdom alongside man. Man, who has broken the chain of instincts, represents the culmination of psychic evolution toward pure intellect. The cockroach, on the other hand, as an insect, represents the culmination of psychic evolution toward pure instinct. In the scale of invertebrates the insect is the crowning point of instinctive life. And the cockroach, being one of the simplest, most ancient and therefore most successful insects, stands on the summit next to man as one of the crowning achievements of life as we know it!

What this means in the practical sense is the following: Just try and get rid of cockroaches! YOU CAN'T. Billions of generations of them have survived essentially unchanged; there is practically nothing that can be done to disturb their balance of life. This applies to all insects. In fact, there are more species of insects known today (900,000) than of all other animals put together (262,500). Pest-control operators (also known as exterminators) know this and they are no fools. Pest-control operators always stipulate in their contracts *control* rather than eradication. Complete eradication is an impossibility.

Just what is it about the cockroach that makes it so successful, so hard to get rid of? Basically, it is its tremendous reproductive capability, its ability to adapt and protect itself, its omnivorous appetite, and the scarcity of effective natural enemies which make for its distressing abundance. Cockroaches and man unfortunately inhabit the same ecological niche. Everywhere people can live, roaches can live, but they can also live in places people can't. Yet man has inhabited every part of the globe. Of the roughly 3,500 known species of cockroaches (of which less than 1 percent are housepests) there are tropical forest cockroaches, semiaquatic varieties, woodborers, kinds that live underground and species that spend their whole lives being parasitic on other insects. However, we are only interested in the ones we see.

The name for cockroach in the following languages is:

Norwegian	*kakerlakk*
Swedish	*mort*
Danish	*kakerlak*
Spanish	*cucaracha*
German	*Küchenschabe*
French	*blatte, cafard*
Dutch	*kakkerlak*
Portuguese	*bicho de-conta*
Polish	*karaluch*
Italian	*blatta*
Russian	*tarakán amerikánski* (American)
	tarakán turusak (German)
Chinese	*chang-lang*
Japanese	*abula mushi*
Hebrew	*juke (jook)*

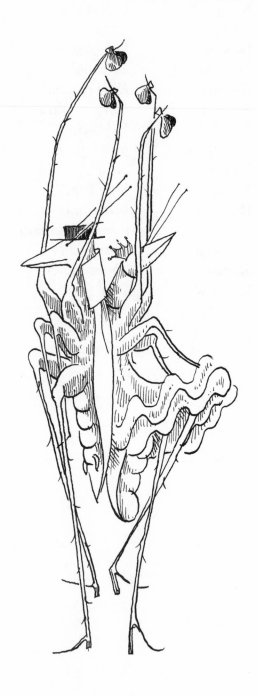

Cucaracha—The Spanish Cockroach

Küchenschabe—The German Cockroach

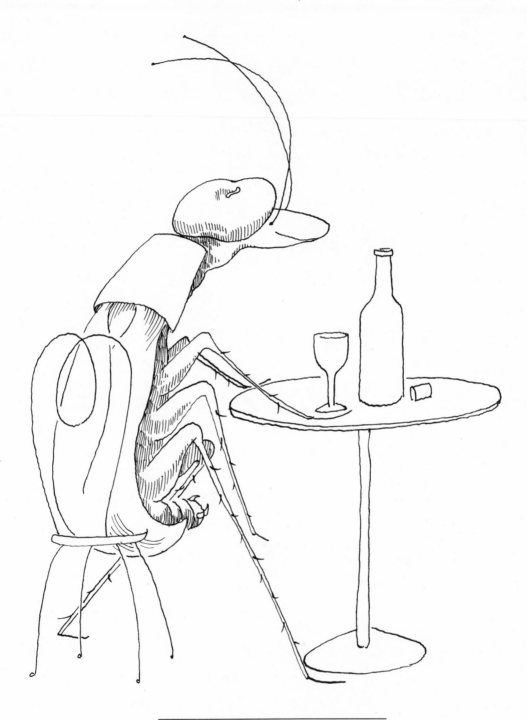

Blatte, Calard—The French Cockroach

Kakkerlak—The Dutch Cockroach

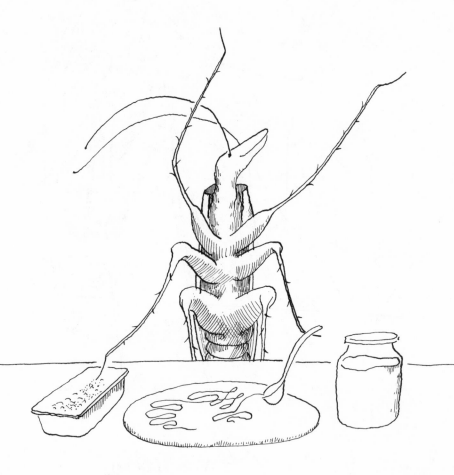

Blatta—The Italian Cockroach

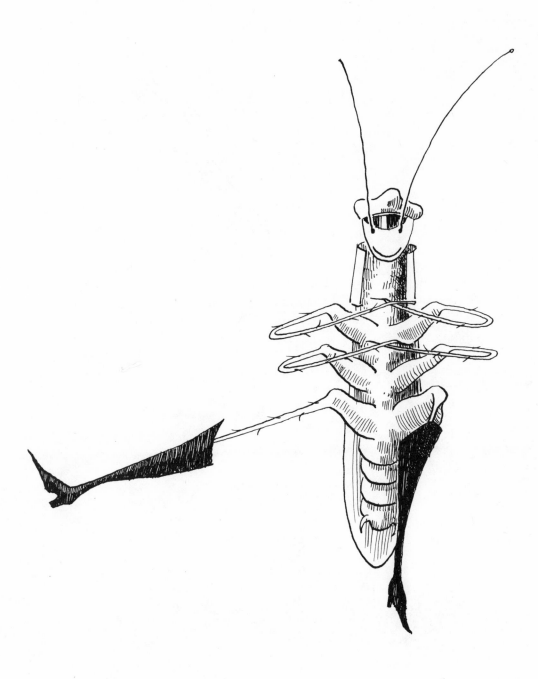

Tarakán—The Russian Cockroach

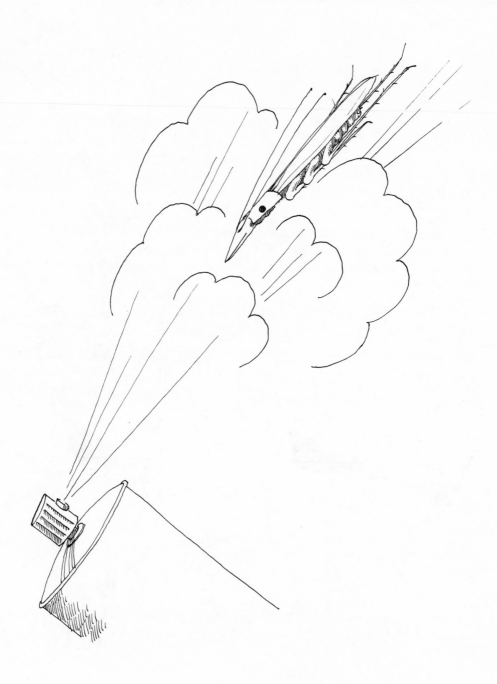

Abula Mŭshi—The Japanese Cockroach

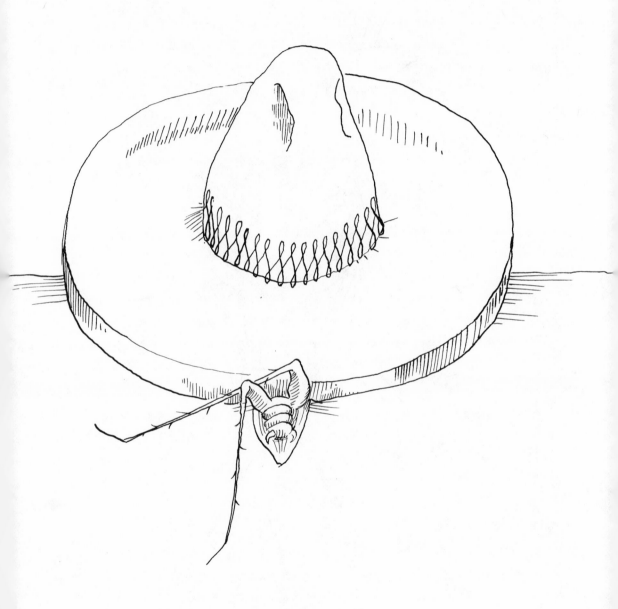

Cucaracha—The Mexican Cockroach

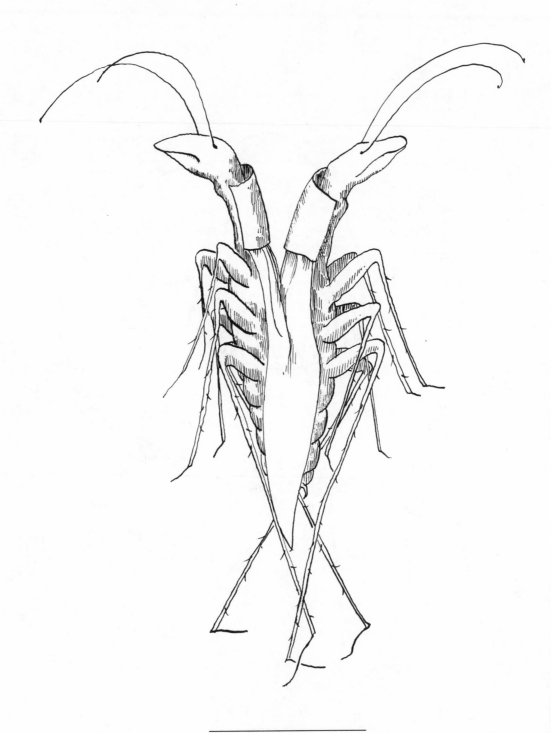

The Siamese Cockroach(es)

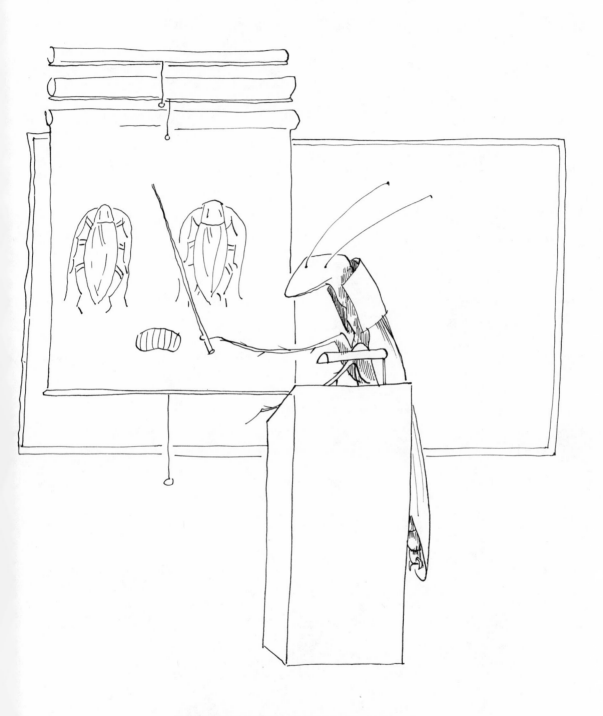

Cockroach lecturing with aid of chart

What do cockroaches eat?

Answer: Almost anything, and it doesn't take too much to keep them happily fed. A thin layer of grease on a piece of steel is sufficient. It is virtually impossible to starve a roach in locations where people live or process food. Cockroaches will eat bread, fruit, crackers, grease, sweets, vegetables, dog food, garbage, beer, marijuana, cigarettes and cereal, to name just a few roach delicacies. They will even eat paper, although we do not know how much nourishment they derive from this. In the tropics it is not uncommon for cockroaches to feed on fingernails, toenails, pus sores and wound openings. If roaches are confined to a cage, they will become cannibalistic and eat each other.

How long can a cockroach go without eating?

Answer: The bigger the roach, the longer it can go without eating. The larger American roach can go three weeks without food or water and over a month with just water. German cockroaches are smaller and can last about two weeks without food or water.

What is a water bug?

Answer: A true "water bug" is not a cockroach but a different insect species belonging to a totally different order of insect. Generally when people use the term "water bug," they are referring to the largest cockroach they may have seen, leaving the definition in the eyes of the beholder. The larger roaches found in basements are American or Oriental roaches, which reach up to

You are what you eat.

three times the size of the German ones. They are found in basements or on city sidewalks because they commonly dwell in sewers and will often migrate when the sewers become blocked up (such as after a heavy rain). They like darkness, and basements are ideal locations for them. With the advent of compactors the American and Oriental roaches are finding easy access to buildings by using the compactor chutes.

What is a palmetto bug?

Answer: A Palmetto bug is an American cockroach found in the southern part of the United States, particularly Florida. It lives indoors and outdoors and is a good flyer. It will wing its way from the outdoors to the inside of a house. In the North the same insect confines itself to be indoors and does not fly. However, recently we have found a few situations where in areas of extreme heat, the American cockroach will fly. We do not know what makes it fly in the South and not in the North.

Why do cockroaches come out at night and where do they go during the day?

Answer: Through evolution, those cockroaches which were active at night survived and those which were active during the day perished from predators. Cockroaches have developed negative phototropism (which means they seek out darkness and try to avoid light). During the day they hide in cracks and crevices.

Where do cockroaches live?

Answer: Cockroaches prefer to form pockets together in locations close to food and water and where it is dark. The German cockroach prefers a feeling of tightness and therefore favors smaller cracks and crevices, while American cockroaches sometimes hang on the wall in the dark. The southern American cockroach (the Palmetto bug) hides in the dark areas of plants, such as under leaves. German cockroaches also prefer to group together on surfaces like wood, cardboard or paper (for example, wall calendars) that their fecal matter will adhere to. Their feces appear as small black specks, which are a good indication of their presence.

Is there such a thing as an albino cockroach?

Answer: No, but what you will see quite often is a white cockroach which has just molted (shed its skin). Cockroaches literally grow by leaps and bounds. They must shed their skeleton, which covers their body. The German cockroach molts six or seven times during the first two months of its life, and the American cockroach molts nine to thirteen times during its first 400 days. When cockroaches shed their skins, they will appear white or albino. Although very tiny and difficult to see or locate, the newly emerged baby cockroaches (called nymphs) are white. Within several hours they will darken.

Do cockroaches protect their young?

Answer: No. They have no parental instincts, and the female German cockroach is the only one that will carry its egg capsule until it is ready to hatch. Brown-banded cockroaches are meticulous as to where they place their egg capsules and will securely glue them to the backs of furniture, bamboo curtains, shoe boxes and other hidden areas. The American and Oriental cockroaches drop their capsules shortly after they are formed.

Do cockroaches eat their young?

Answer: Normally they don't but will if they are hungry enough.

What do baby cockroaches look like?

Answer: We refer to young cockroaches as nymphs, and these nymphs go through a series of molts. At each stage the nymph's shape stays the same except that it becomes larger. All nymphs lack wings. The color pattern varies with the species. The German cockroach nymphs will pass through either a reddish-brown phase or a more common black phase. In either case they possess a yellow area running down the center of their backs. It is important to be able to recognize an immature cockroach since you do not want to wait until they are adults and reproducing before you undertake control methods. Because the average person is generally accustomed only to recognizing the adult cockroach, he or she often mistakes the nymph for another type of insect.

English roaches have adapted admirably to their environment.

Why do cockroaches group together?

Answer: In the laboratory we have been able to show that when cockroaches are isolated in individual chambers, the rate of survival is lower and the time to reach adulthood is longer. The benefit of a group effect can only be speculated on. However, it is suggested that by grouping together, cockroaches create a "micro-environment" most suitable for their survival in terms of maintaining an ideal relative humidity.

Is it possible for me to get rid of cockroaches in an apartment if my neighbors have them?

Answer: Yes and no. Yes, if your neighbors cooperate and allow all apartments to be treated properly. No, if at least one neighbor does nothing and allows the cockroaches to breed, run along the water pipes and be thrown out in the garbage.

What can I do to decrease the number of cockroaches coming from a neighbor?

Answer: Stuff steel wool or other flexible material into holes around pipes from adjoining walls, floors and ceilings. Try to get your neighbor to cooperate with the professional pest-control technician contracted to do the apartment building (if such a service exists).

Is the presence of cockroaches in my place a sign of poor housekeeping?

Answer: No. A few cockroaches can be brought in undetected almost anytime. However, the buildup of a large population occurs from neglecting the problem and can be considered a sign of indifference.

Is it ever possible to get rid of roaches completely in a home?

Answer: Yes. However, you may have a reinfestation if you live in an area where an influx of new cockroaches is an almost daily occurrence. If you are having trouble with roaches, you will be better off to rely upon your professional pest-control operator.

What are cockroach droppings?

Answer: Cockroaches plaster their fecal matter to wooden or cardboard surfaces. In areas where the fecal matter does not stick, it will drop to the nearest surface. The scientific analysis of such droppings is called scatology. Each cockroach species produces a uniquely shaped and sized fecal pellet.

How smart are cockroaches? Do they have brains?

Answer: Cockroaches do not have brains in the normal sense but do have a concentration of nerves in their heads which serve as the focal point of their nervous systems. Cockroaches can learn by experience how to travel mazes. They also have enough "sense" to try to avoid treated pesticide surfaces and move rapidly from extremely hot, dry or cold areas. This creature has been able to survive for 350 million years, and despite man's efforts to eradicate it, it continues to thrive. Biologists view the survival instincts of the cockroach with high regard.

Can cockroaches eat toothpaste or soap?

Answer: Cockroaches will eat toothpaste. We know that cockroaches will eat soap, but we do not know if they will digest it. I might point out that mice readily feed on soap, probably for the fat content.

I left my coffee cup on the kitchen counter half-filled with coffee. The following morning I found a dead roach floating in it. How did it get there?

Answer: The cockroach's antennae were able to detect the odor and moisture emitted from the surface of the coffee. The roach climbed into the cup in search of food and/or water and failed to get a good foothold. He fell in, could not get out and drowned. Cockroaches can swim, but eventually they must be able to pull themselves up onto "dry land."

Last night I ordered a drink in a restaurant and noticed a cockroach petrified inside an ice cube. How did it get there?

Answer: Ice cubes are made by freezing water in machines. Cockroaches are naturally attracted to wet areas. In this case the roach slipped into the water.

What can I do to decrease the chances of getting cockroaches in my house?

Answer: First, make sure to check all paper bags, cardboard boxes and laundry bags before bringing them into your home. If you find *one* cockroach, do *not* bring that bag or carton into the house. You can take the contents out and bring them in, but make certain to search each item. Second, in apartment houses, if there are holes in the wall around the pipes, plug them up with steel wool until you can have them permanently sealed. Avoid stuffing paper bags between cabinets and the refrigerator because these areas harbor roaches and do not come in contact with the insecticides that were placed in cracks and crevices. Throw out your garbage daily. Finally, have your premises treated regularly.

What good are cockroaches?

Answer: They are natural scavengers and they help clean up waste materials. However, our domestic species currently serve no known beneficial purpose.

What do I do if my guests arrive and a cockroach appears?

Answer: You can do three things: (1) Don't spray since you will only flush more roaches out because of the irritant in the chemical formulation (besides, your guests will smell the odor of the spray and know you have roaches); (2) leave the lights on because cockroaches prefer to stay in the dark; and (3) stifle yourself—don't tell your guests and call your pest-control operator the next day.

What is insect phobia?

Answer: Insect phobia is an exaggerated fear of the presence of nothing. The insect-phobic person believes that there are insects present, but there is nothing there. These individuals will go so far as to develop rashes, lose sleep and come close to a nervous breakdown. Many times the cause of this condition is an allergy.

Do cockroaches bite people?

Answer: Cockroaches have chewing mouth parts, and, in some cases where there are heavy infestations, we have seen them feed on human sores, remove eyelashes and nibble on fingernails and toenails. But these situations are not everyday events. They occur only if the cockroaches are allowed to build up over a long period of time.

What are the control methods of the future?

Answer: We look to a pest-management approach whereby sanitation and the elimination of favorable roach harborages will play a major role and pesticides a minor one. We also look forward to chemical formulations that will interfere with the roach's prolific reproduction process.

What purposes do the odors of cockroaches serve?

Answer: There is a distinct variation in the odor of different species. A good roach man can walk into an apartment, take a good whiff, and if there is a heavy population present, he can identify the species. These odors serve sexual purposes and encourage pocketing.

How long do cockroaches live?

Answer: German cockroaches can live up to a year. American and Oriental cockroaches can survive for two years or longer.

What is the difference between licensing and certification for pest-control operators?

Answer: A pest-control operator can obtain a license by paying a fee, so his having a license does not necessarily mean that he is qualified. Certification, on the other hand, means being competent in that the pest-control operator has passed a written examination. By 1978 most states should have certifying examinations.

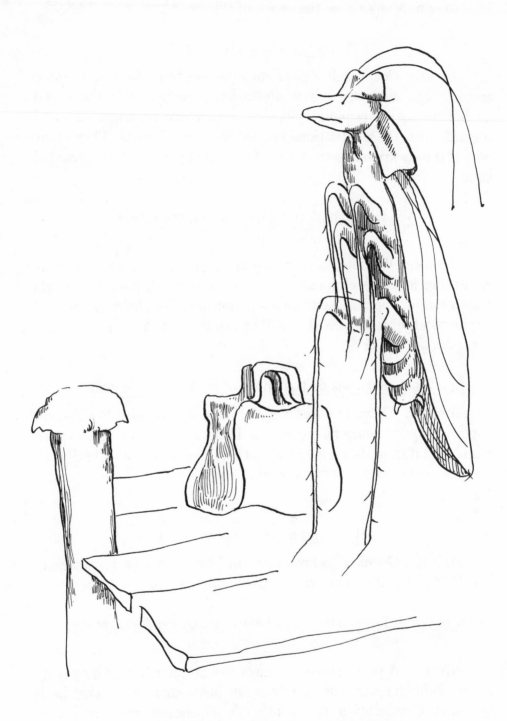

German Cockroach having just landed at Ellis Island.

Do Chinese restaurants have more roaches than other kinds of restaurants?

Answer: No. Clean restaurant managers will keep their restaurants pest-free. Careless managers, regardless of race or color, will encourage roach populations. Oriental roaches came originally from the tropics.

Is it true that cockroaches can withstand atomic radiation?

Answer: Cockroaches are able to survive exposure to radiation levels far in excess of what man can withstand.

Are certain restaurants more likely to have cockroaches?

Answer: Definitely. Some restaurants have no pest-control service. Restaurants which hire employees who have cockroaches in their homes may find their workers bringing them to work. Restaurants which order food and use vendors who have their own cockroaches are most vulnerable to a continual influx of these creatures.

I thought only dirty homes had roaches?

Answer: Yes and no. People who keep dirty homes may be poor housekeepers and tend to do little when they find cockroaches. Therefore, the roaches persist and thrive. Initially cockroaches are not prejudiced. They will enter anyone's home. Anybody can get cockroaches. It's no disgrace to have cockroaches; it's only a disgrace to keep them.

Why haven't cockroaches developed resistance to boric acid?

Answer: To date, cockroaches, after exposure to various organic synthetic (man-made carbon products) insecticides over many generations, have exhibited genetic resistance to them. These insects develop biochemical pathways to detoxify or neutralize the pesticide. Boric acid is not an organic synthetic. It works on several different systems of the cockroach, thereby making resistance to it more difficult. Boric acid is both a stomach and a contact poison.

What is Bolt?

Answer: Bolt is a trade name associated with an entire line of pesticides marketed by S. C. Johnson & Son, Inc. One product which has given good control of cockroaches in apartment closets is Bolt Roach Bait.

How old are cockroaches?

Answer: Cockroaches have roamed the surface of the earth for about 350 million years. Today's roaches look very much like their ancestors. Fossil records document this.

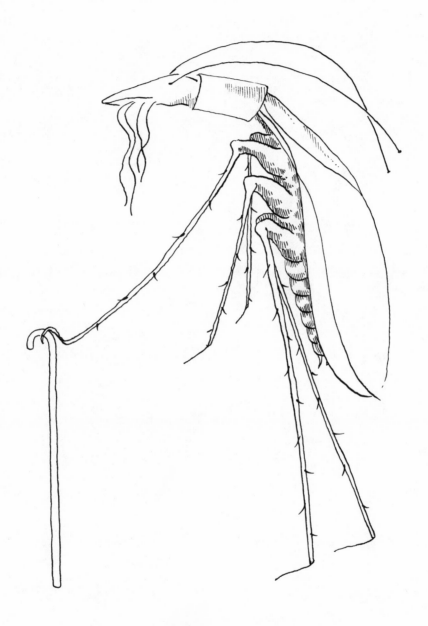

The Oldest Cockroach

Timid roach why be so shy?
We are brothers, thou and I.
In the midnight like yourself
I explore the pantry shelf!
—CHRISTOPHER MORLEY

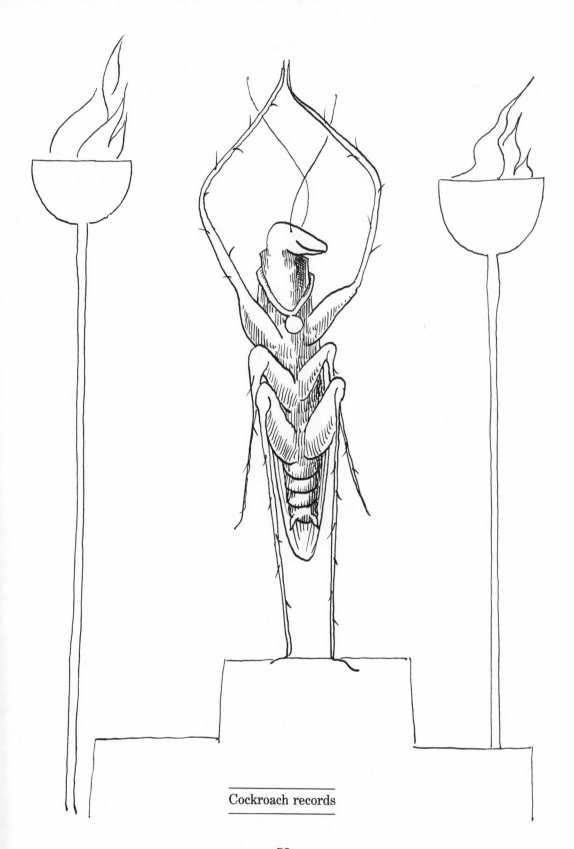

Cockroach records

4. COCKROACH RECORDS AND STORIES

All cockroaches can swim for a short period of time. Of the domestic species, Oriental and American cockroaches are the best swimmers. They are often associated with sewers and can swim through pipes. Some nondomestic cockroaches live in and near water, while others are desert dwellers.

Some species can hiss or whistle by blowing air through their spiracles (small holes on the sides of their bodies). The loudest of such roaches are called *Gromphadorhina* spp. They are sometimes called hissing cockroaches because when one goes to pick them up, they will hiss. The sound of the hissing mimics that of a snake.

Blaberus giganteus, a tropical cockroach, is the largest in the world. It is kept by entomologists in rearing containers, primarily as a conversation piece. These cockroaches live under rocks, and in caves in Panama and palm trees in Trinidad. They measure about three inches in length, not counting their long antennae.

If you care to know the longest distance a cockroach has been scientifically recorded to have moved through sewers, this record goes to the American cockroach. One has been clocked moving a total of 385 yards (almost the distance of four football fields). This record was authenticated by capturing live cockroaches in a sewer in Tyler, Texas, tagging each one as one might a bird, and then releasing them back into the sewer system at the same point where they were captured.

It is interesting to note that before heavy rains, cockroaches dwelling in sewers will migrate in large numbers into basements to avoid drowning. Cockroaches can detect changes in barometric pressures and therefore know when to take to the high ground. In

Which cockroach is the best swimmer?

Bermuda, thousands of cockroaches have been observed running across the highway about an hour before a downpour. The native Bermudians recognize this as a signal of forthcoming rain.

Cockroaches can hide or scurry into the smallest places imaginable. The newborn cockroach can hide in a crevice one-half of a millimeter wide (which is as narrow as a thin piece of paper). A hungry, adult male can fit through a crack thinner than a dime.

An entrepreneur in Long Island, New York, copyrighted (under Super Zoom Film, Co.) and marketed a black-on-green bumper sticker reading "Warning! I Brake For Cockroaches."

The Velsicol Chemical Company developed a fictitious character dubbed "Rudy the Roach" as part of a national sales campaign. This company markets T-shirts and buttons emblazoned with the cartoon Rudy. A few toy companies have experimented with rubber cockroaches. Some pest-control companies adorn their service vehicles with four-foot cockroach models.

There is a story current among pest-control operators that a woman in Miami, Florida, was seen at a supermarket leading a large cockroach pet on a leash. However, to date this story is unsubstantiated. Some pest-control operators have even been assaulted or harassed by apartment dwellers who have developed a personal liking for their "pets" and have feared that they might be exterminated. Pet cockroaches are inadvisable because of their ability to transmit disease organisms.

In 1965, in Lafayette, Indiana, 6,000 cockroaches were found in one beer carton.

On August 1, 1947, a four-room apartment in Austin, Texas, was treated for German cockroaches. Almost 100,000 cockroaches were counted killed that day.

In 1967 when Stanley Rachesky, the entomologist and author, tested the effects of the pesticide Baygon, he collected 132½ pounds of dead American cockroaches from the cage of Sinbad the Gorilla at the Lincoln Park Zoo in Chicago.

In the 1880's, according to Army records unearthed by Smithsonian Institution researchers, a brigade of men wielding brooms were unable to halt a mass migration of roaches out of a Washington, D.C., restaurant and into the U.S. Capitol. Today, "we can't seem to get rid of them," stated a despairing official in the Capitol architect's office.

In 1977 three New York University students made news when they bought a foot-long gecko lizard to eat the cockroaches in their Greenwich Village apartment. "We had so many cockroaches the kitchen sink was black at night," complained one student. After several months the lizard began to get the upper hand in the control of the problem. "We used to hear him crunching on them at night. It woke us up at first, but after a few nights we got used to it," another of the students added. (This story was reported in *The New York Times*.)

In May 1938, a prisoner in the Amarillo, Texas, jail told how he had trained a cockroach to come to his solitary-confinement cell when he whistled. The cockroach would come with a cigarette tied to its back.

Cockroaches are closely related to termites. In Australia there is a termite that deposits eggs in a capsule similar to the one produced by some cockroach species.

In 1969, in Woodbridge Township, New Jersey, it was reported that hordes of cockroaches invaded homes during the night in search of food. "It's like something in the movies," said a distressed homeowner whose house was one of the ten severely hit by the insects. "Armies of them march across the street every night," she stated.

Peter C. Sheretz, a teacher at Virginia Commonwealth University, Richmond, says that the hardy cockroach can dine on lethal cancer-causing toxins without any harmful effects. He expressed the hope that this finding may be helpful in discovering a cure for cancer. "I fed them higher concentrations of the cancer-causing agent than a lot of other researchers have fed to other animals which got cancer," Sheretz said. "They ate it until they really died of old age."

On December 13, 1976, the *Long Island Press* reported that Bruce Hammock, an insect endocrinologist, kept 500 giant Madagascar and Panamanian cockroaches locked in a special escape-proof lab at the University of California at Riverside. "These are the biggest roaches on earth," Hammock stated, fondling the roaches and allowing them to crawl on his lab coat. He is trying to find a synthetic hormone that can replace insecticides.

Women entomologists have been in the forefront in research on cockroaches. At the American Museum of Natural History in New

York City, Dr. Betty Lane Farber has been tagging cockroaches and studying their behavior. She has reported that "male cockroaches stay out later than female ones and that all roaches give off a distinctive scented signal—a pheromone—that attracts other roaches." She observed that although cockroaches may not have a clear sense of territoriality, they do appear to stay in one place most of their lives. Since most of their activity is at night, Dr. Farber often sleeps on a cot in her laboratory and wakes to track her tagged roaches with flashlight and camera. (Dr. Farber's observations were reported in *The New York Times*.)

Another noted researcher is Alice Gray, an insect specialist at the American Museum of Natural History. She is well remembered for sponsoring in 1973 the first public cockroach exhibition at the museum called "Roaches Are Here to Stay." She has been remarkable in helping thousands of students to become interested in insects. She believes that roaches should be admired for their beauty and ability to adapt.

Dr. Ruth Simon, a member of the U.S. Geological Survey, observed that cockroaches can predict earthquakes. "Before an earthquake of small intensity, there's a marked increase in activity," she reports.

Research conducted in 1977 by Genie Floyd, a freshman at New College, Sarasota, Florida, indicates that cold cockroaches are smarter than warm ones. This marine-biology major found that cockroaches that had been subjected to a temperature of 43° F. made fewer errors when escaping from a maze than did cockroaches kept at 73° F.

Dr. F. A. McKittrick is another noted woman entomologist. She specializes in the study of the evolution of cockroaches and has published the major work on this topic.

Women have also been entering the field of professional exterminating. In New York City two women recently started a company called "Lady Bug Pest Control," and another woman has entitled her business "The Elegant Exterminator."

Cockroaches are known to prevent transistor radios from working. Roaches have chewing mouth parts with teeth like mandibles that are capable of destroying wires and connections in electrical appliances.

Dr. Frishman encountered an extremely heavy population of

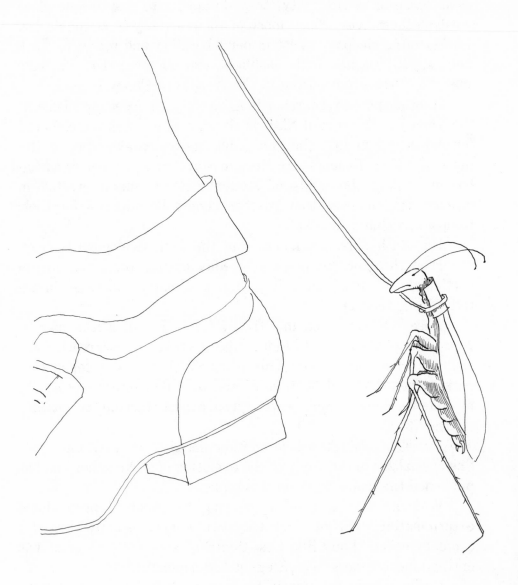

A pet cockroach

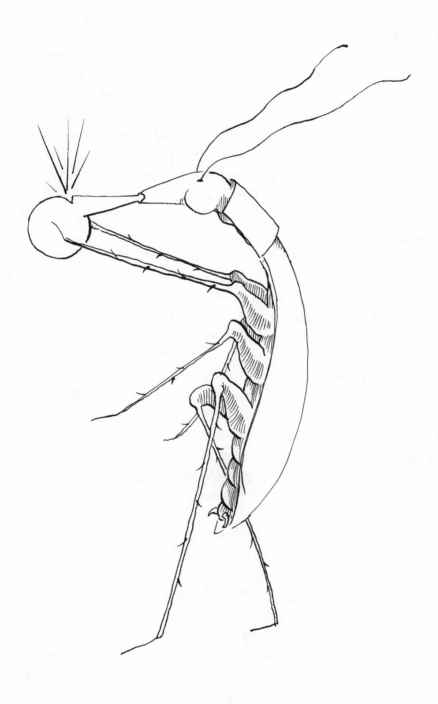

What cockroach can whistle the loudest?

cockroaches living inside a console radio at a major city hospital. There were so many of them that it was difficult to read the markings for the stations. The cockroaches were clustered just under the plastic protecting the dials. Thousands more were hidden behind the backplate. In order for the cockroaches to survive, they would rush out at night and drink urine, which had been deposited on the floor by patients who could not control themselves. Dr. Frishman had been brought in to deal specifically with this problem. His solution was simple. He placed a large plastic bag over the entire console unit and put a pyrethrin bomb inside to destroy the population. A few hours later he vacuumed up the roaches, threw away the bomb and restored the console for the entertainment of the patients.

The Callithricid monkey removes the head, guts, wings and legs of a cockroach before eating it. In some cultures man also eats cockroaches and their egg capsules. Cockroach-eating cultures have been found in Australia, Japan, Thailand, China and other areas of the Far East. However, it is strongly advised not to become a cockroach gourmet because the oil secretions of the roach, its bacteria and other fauna could be harmful.

Cockroaches are found throughout the world. In 1977 *The New York Times* reported that among the products most sought after by the diplomatic community in Moscow was cockroach spray. According to Alice Gray of the American Museum of Natural History, few cultures accord the cockroach any reverence. Only in India and Polynesia are jewelry and ornaments devoted to the roach to be found.

The largest number of cockroaches emerging from a single egg capsule is fifty-two nymphs. Normally the German cockroach produces about forty young per capsule for her first three or four capsules. Thereafter the numbers decrease. The record of fifty-two nymphs was personally reported by Dr. Frishman.

The New York Times reported on November 20, 1978, that despite the pesticide bombing of 4,500 Metropolitan Authority buses every other weekend, this has failed to control cockroaches living in the seats, walls and floors of the buses.

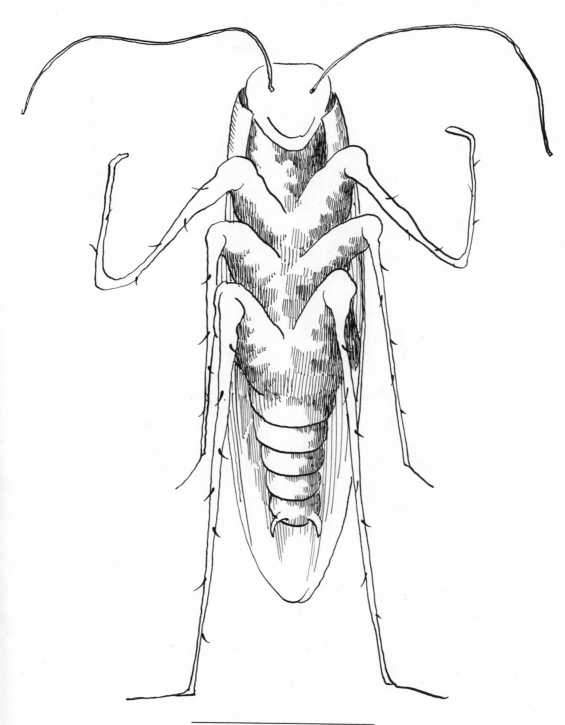

Which is the largest cockroach?

Cockroaches have been known to fly.

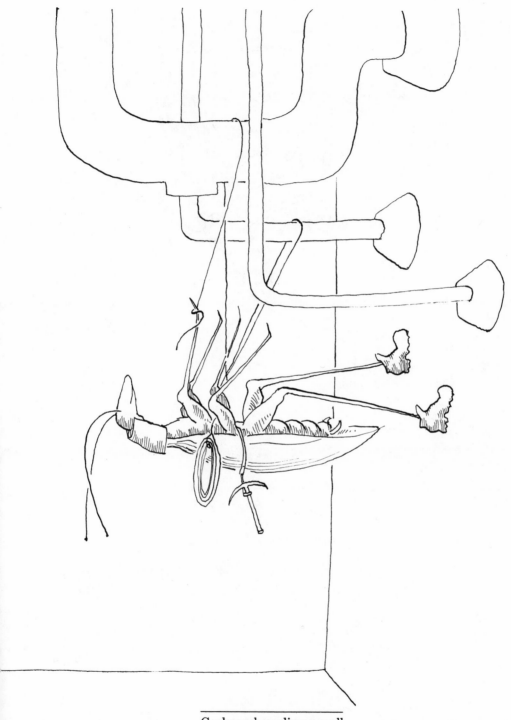

Cockroach scaling a wall

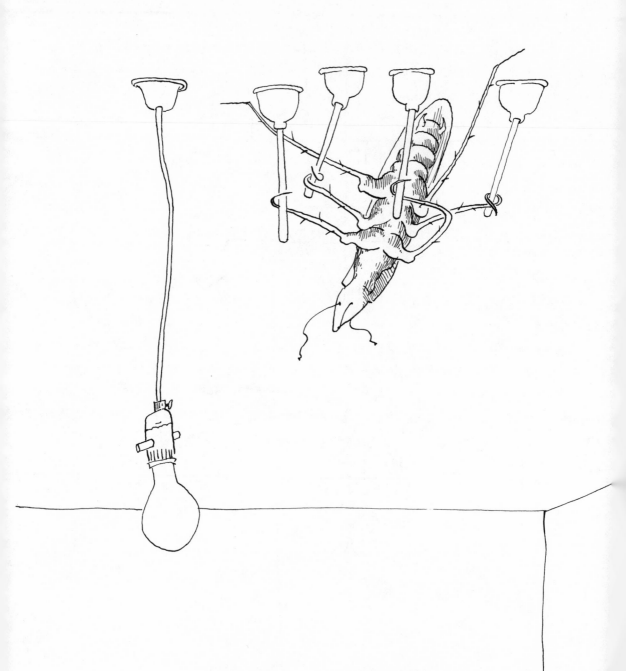

Cockroach carefully traversing your ceiling

Person about to smoke a roach

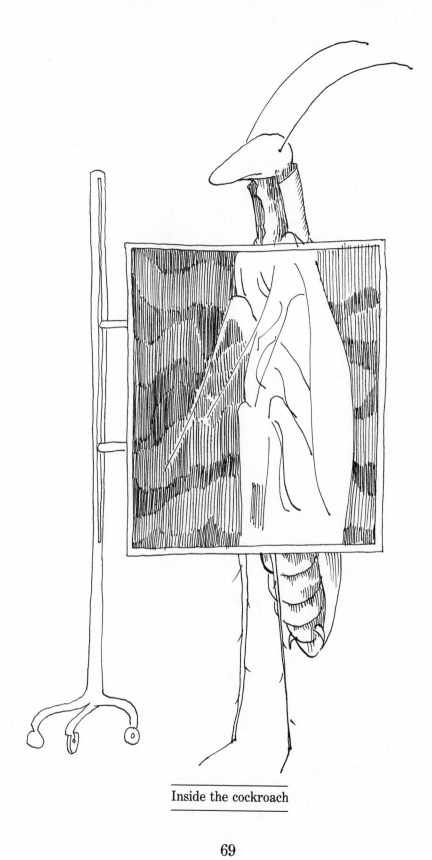

Inside the cockroach

5. THE BIOLOGY OF THE COCKROACH

Fossil cockroaches show no great differences from present-day species, and today's cockroaches doubtless live much as their ancestors did over a billion generations ago. The fact that they still exist after so long a time, let alone that they are basically unchanged, is a magnificent tribute to the success of their structure of life. The cockroach is a grandly simple creature, a living fossil, unencumbered by sophisticated evolutionary amenities such as lungs, reliance on vision and discriminatory taste.

The basic domestic cockroach ranges in color from yellowish tan to brownish black and in size from one-half inch to one and one-half inches. It is not slimy, but is covered with a hard, waxy coating. Like all insects, it has a three-part body. However, you never see the three parts, not even the head, unless the cockroach is dead and happens to be lying on its back. This is because its wings cover the entire back and its head is bowed downward under a protective crown with the mouth projecting backwards between its front legs. Therefore, you mainly see the back, legs and antennae sticking out, and maybe the cerci, or feelers, on the rear end.

The cockroach does most of its sensing by picking up vibrations in the air, essentially hearing or feeling movements by means of its antennae on one end and the cerci on the other. The antennae can also smell things. The cerci are directly connected to its six long, slender, powerful legs, bypassing the brain. When the cockroach "feels" a presence anywhere in the room (when you speak a word, take a breath or take a step) it doesn't even think about it, its legs start running first. Evolution has favored the shortening of evasive

71

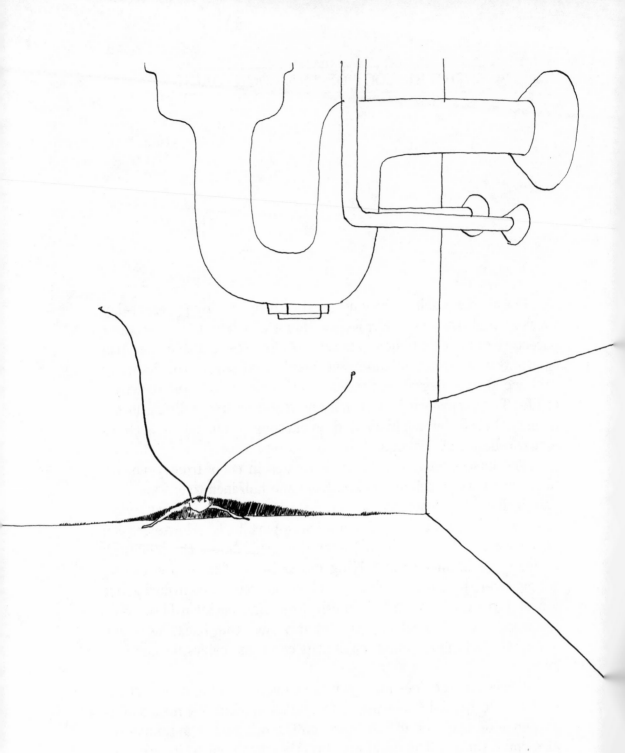

Roaches like sinks.

behavior responses in the cockroach to the point where its reaction time is limited only by the speed of neural impulses down the nerves from the cerci to the legs (this reaction time has been measured). If you ever observe a cockroach standing still with its cerci sticking up in the air, you'll know it's checking things out. There is no way you can get near a cockroach unless it starts running and can't find a place to hide. This is a rarity for a roach because its body is built of flattened plates with the legs sticking out horizontally, and it can slip into cracks you couldn't get a match into. It will also use its wings if properly disturbed.

A cockroach needs very little to survive. A little warmth, a little shelter, a little moisture and a very little amount of food. Although it prefers starchy food like bread, potatoes, apples and beer, it will eat anything at all. The cockroach is not capable of biting; rather, it scrapes and chews or munches. It will just as soon snack on paper and soiled clothing as on scattered crumbs. It seeks high humidity and warm temperatures. It tries to avoid light and prefers to hide in cracks and crevices. A cockroach is extraordinarily hardy. It will survive the loss of legs or antennae. It will withstand temperatures from 10° F. to 140° F. for short periods, in addition to depressurization. It can go for long periods of time without food or water. While it lives and breeds in close proximity to its food supply, it will wander or forage and even migrate if it has to in search of food. But it really doesn't take much to keep any cockroach fat and happy.

Cockroaches come in four flavors: male, female, nymph and egg. The male is thin and slender with a tapered rear end, whereas the female is stout and robust with bulbous hindquarters. The nymph is simply the stage between hatching and maturity.

Cockroaches are highly prolific; the female is essentially a fertility machine. Although times vary among species and environmental factors play a part, it is just a matter of days between reaching maturity and producing the first batch of eggs. Cockroaches need only mate once in their lives to produce many batches of fertile offspring; however, they are known to do it again and again anyway. Incidentally, some can and will reproduce parthenogenetically; that is, without any male intervention. Such eggs laid by females give rise to new generations consisting of only females. A word here for cockroach lib?

Feeding your cockroach

An expectant roach

Cockroaches always hatch from egg capsules, little lima bean-shaped cases in which baby cockroaches are stacked like coins. In some species the female carries the egg capsule protruding from her posterior until the babies are ready to hatch, while others deposit the capsule as soon as it is formed. Spotting an egg capsule is worse than spotting a real live cockroach because with the egg capsule, you are looking at sixteen to forty potential real live cockroaches.

When cockroaches hatch, they are nymphs. The only difference between a nymph and an adult is that the nymph doesn't have wings or a few other things you can't see, such as genitalia. The nymph also does not have the distinct markings that the adult has.

The way a nymph becomes an adult is by molting. Molting means shedding one skin for a new one. A nymph molts by breathing in until it bursts its skin, which also happens to be its exoskeleton. The exoskeleton splits right down the middle, and the nymph squirms out, this time a little larger. Nymphs molt from six to a dozen times before they become adults. If a nymph should happen to lose any appendage during this time, it will just generate another one at the next molt. Any stories about albino cockroaches you've heard are simply stories about newly molted nymphs. They shortly turn dark with exposure to air. Don't feel bad if you've been fooled by these white nymphs because apparently their fellow cockroaches don't recognize them either—the nymphs are highly susceptible to cannibalism until they darken.

Description unnecessary—Authors

White nymph being prepared

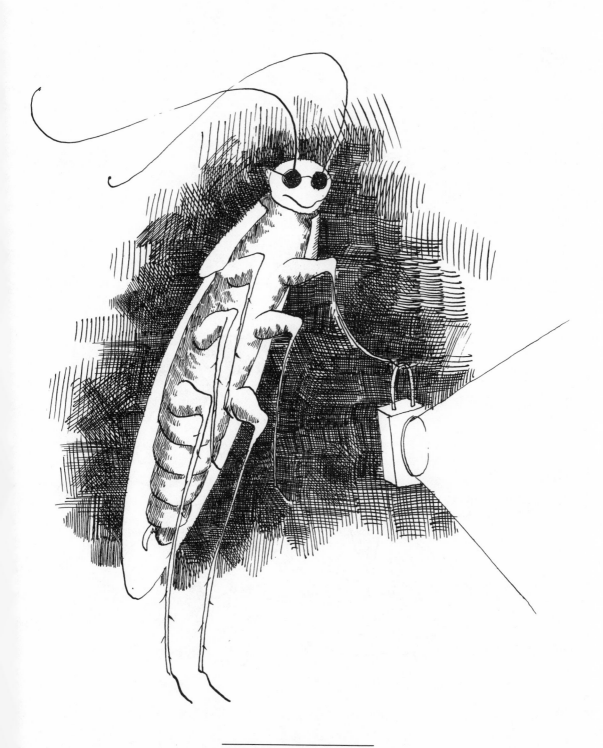

Cockroaches like darkness

Cockroaches enjoy the warmth.

Cockroaches like it damp.

Profiles of the four most common species

German Cockroach *(Blattella germanica)*

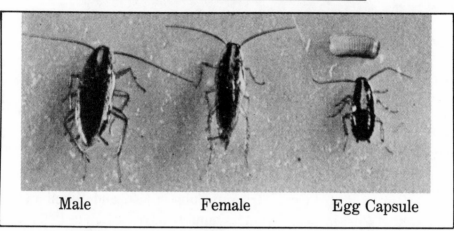

| Male | Female | Egg Capsule |

Nicknames: Steam Fly, Croton Bug. (Large numbers of this cockroach appeared in New York City shortly after the completion of the Croton Aqueduct, which supplies New York with water. Hence a possible origin for the name; however, "Kroton" in Greek means tick or bug.)

Size: ½″ to ⅝″. *Color:* Light brown (tawny).

Markings: Two dark streaks running lengthwise down the back.

Color of egg capsules: Yellowish or red-brown. Always carries egg capsule until ready for hatching.

Number of eggs per capsule: 30 to 48.

Young can mature in: 36 days.

Annual descendants: 400,000.

Where found: Generally thrive in cooking areas. German cockroaches love heat and moisture. They abound near hot-water pipes, under moist sinks, in stoves and hotplates and behind refrigerators.

Brown-banded Cockroach *(Supella longipalpa)*

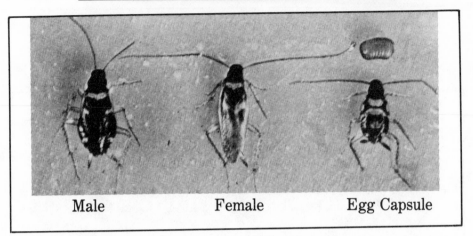

Male   Female   Egg Capsule

Nicknames: Brown Bandit. (Originally introduced to the United States from Africa to Cuba in 1892 and Miami and Key West in 1903. From its humble introduction in 1903, it rapidly spread across the country. In 1967 it was reported in every state except Vermont. The rapid spread of this species across the United States since its introduction this century is probably attributable to its habit of hiding itself and its egg capsules in luggage and in furniture.)

Size: ⅜″ to ½″. *Color:* Dark brown to pale golden.

Markings: Two transverse brown bands across the back.

Color of egg capsule: Yellowish or red-brown. Deposits egg capsules in out-of-the-way places like the undersides of shelves, on the bottoms of drawers, in book bindings and TV sets.

Number of eggs per capsule: 18.

Young can mature in: 54 days.

Annual descendants: 135,000.

Where found: Are generally found in bedrooms, closets and in piles of clothes. Prefer locations high up in heated rooms, including closet shelves, desks and bureau drawers, inside book bindings, behind pictures and wallpaper, and even inside books and telephones. Are generally found in clusters.

Oriental Cockroach (Blatta orientalis)

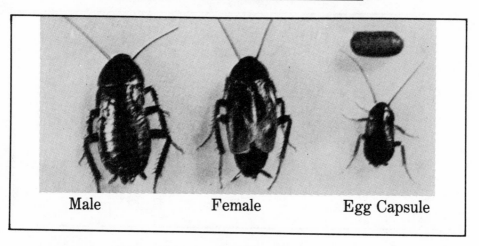

Male   Female   Egg Capsule

Nickname: Water Bug, Black Beetle, Shad Roach. (In Philadelphia, its spring appearance coincides with the arrival of spawning shad fish in the Delaware River.)

Size: 1¼". *Color:* Nearly black.

Markings: Short wings.

Color of egg capsule: Nearly black. Deposits egg capsule.

Number of eggs per capsule: 16.

Young can mature in: 128 days.

Annual descendants: 200.

Where found: Found in cool, damp areas such as basements, service ducts and crawl spaces. Live in sewers and sometimes enter the home through sewage drains. Also found around toilets and baths and sinks, where large numbers congregate around sources of water.

American Cockroach (*Periplaneta americana*)

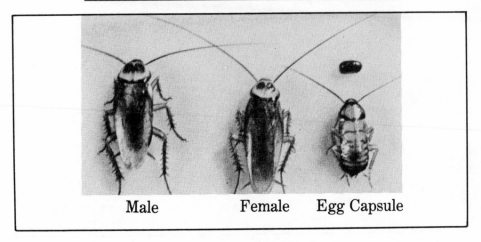

Male   Female   Egg Capsule

Nicknames: Palmetto Bug, Bombay Canary, Water Bug (New York City).

Size: 1½ ". *Color:* Reddish brown.

Markings: None. It's just enormous.

Color of egg capsule: Reddish brown or black. Deposits egg capsules.

Number of eggs per capsule: 16.

Young can mature in: 160 days.

Annual descendants: 800.

Where found: Prefer basements, especially around pipes and plumbing fixtures. Are frequently found in bathrooms and in the vicinity of sewer drains.

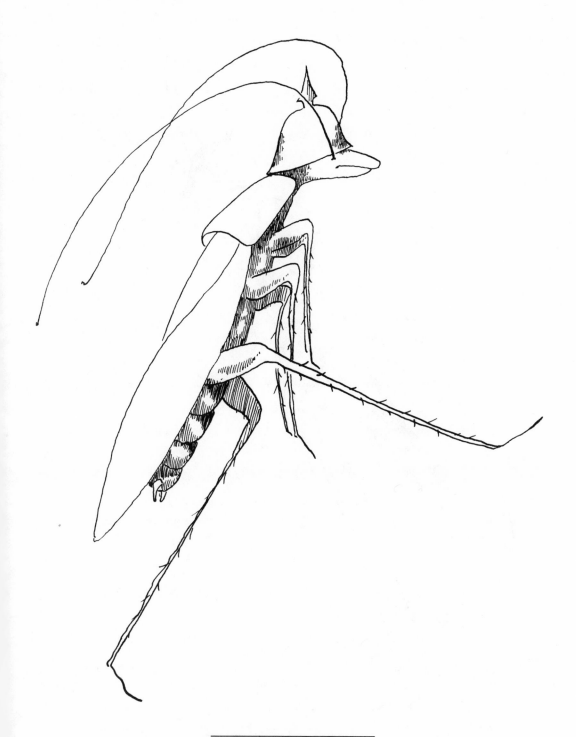

German cockroach marching

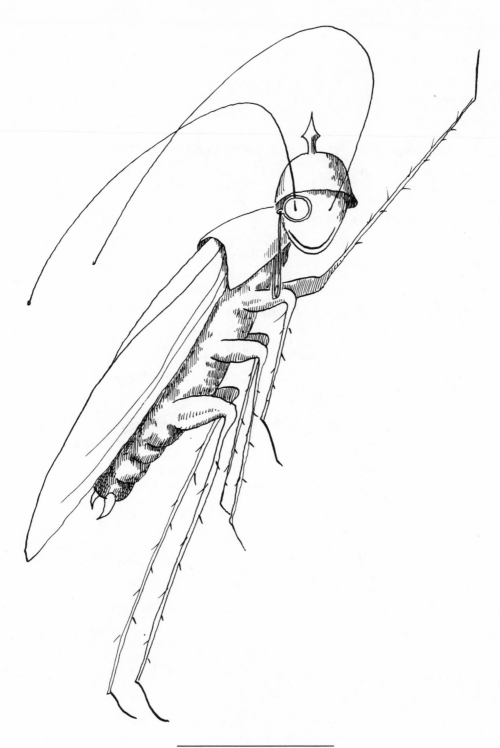

German cockroach saluting

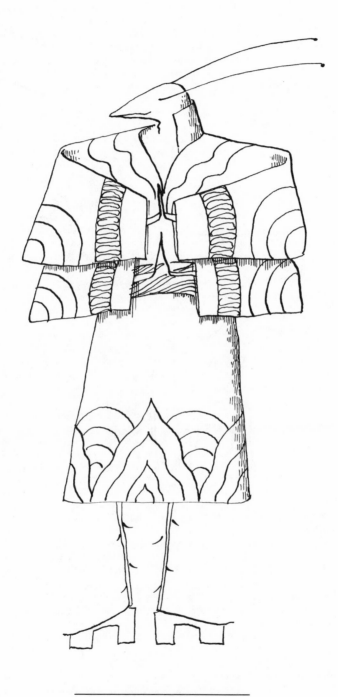

Venerable Oriental Cockroach

Oriental cockroach eating chow mein

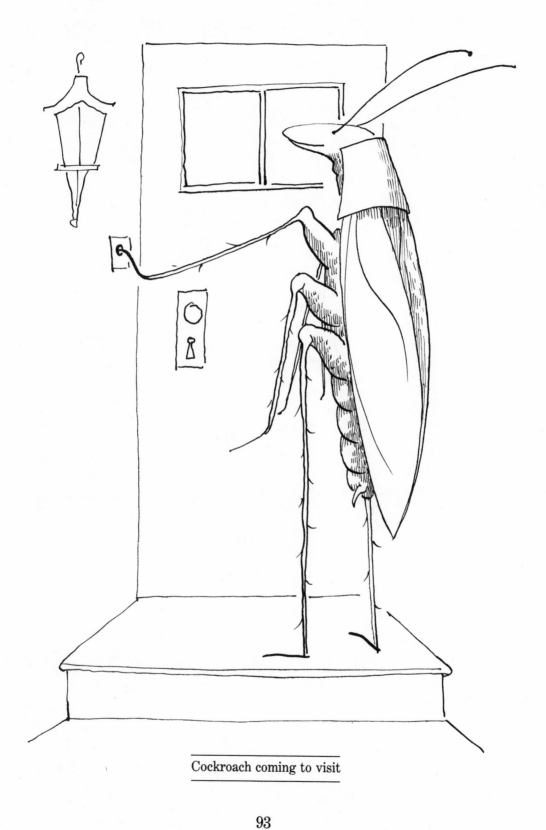

Cockroach coming to visit

At eight-thirty in the evening the telephone rang.

"Are you Doctor Frishman, the bug doctor?" a woman's voice asked.

"Yes. What's the problem?"

Excitedly, she began, "Thank goodness! You're the man I want to talk to.

"I came home from work last night and turned the lights on, and there were a bunch of water bugs by my sink and on the sponge.

"But I want you to know I keep a very clean house. Is there any possibility that they came from my neighbor's apartment?"

I avoided her series of questions and attempted to calm her down.

"There is no disgrace in having insects; it's only a disgrace to keep them." Over the years I had learned that if I told the caller these water bugs were, in fact, cockroaches, her embarrassment would force her not only to deny their existence but to try to convince me that she was correct and I was wrong.

"Well . . . how do *we* get rid of them?"

"For every one you have seen in the sink, there are probably more back in the cracks. So, if you want to do a good job eliminating the roaches, it's going to take some work."

"Okay. What do I have to do? I want you to know that I have already sprayed and it didn't seem to do much good."

"That's because you're not putting the pesticide where the roaches are."

Think like a roach.

(Even though I work closely with pest-control operators, I usually try to give the homeowners a chance to try to control the problem themselves. If I were to tell them that in the long run they would probably have to hire a professional exterminator, they would resent the implication that they could not handle this problem themselves)

"You have to think like a roach. You have to put the pesticide into the cracks and crevices where they are hiding. Now, to do this you have to take out all your pots, dishes, silverware and food and pile them up on top of the dining room table. Then, you have to do the same thing in the bathroom and remove all items from the medicine chest and from under the sink. Now you treat these areas with a residual spray."

"What kind of spray?"

"It doesn't matter whose spray you use [I usually don't like to recommend a specific spray because no one spray is that much better than any other] as long as it contains one of the following chemicals. If you'll get a piece of paper, I'll spell the names for you.

"D-I-A-Z-I-N-O-N ... B-A-Y-G-O-N ... D-U-R-S-B-A-N. These are the major pesticides used in the pest-control industry. There are other insecticides, but as long as you use one pesticide which contains one of these three chemicals and you put it where the roaches hide, then it should work.

"Do not spray in the bread box or on toothbrushes. Let the pesticide dry for about ninety minutes before you put anything back in your cabinet, and *don't* wash off what you just sprayed."

My caller sighs, "It sounds like a lot of work. Isn't there an easier way?"

"Well, the easy way didn't work, did it? And you may eventually have to call a professional pest-control operator."

A pregnant silence follows.

"But, Dr. Frishman, I keep such a clean house. How did the cockroaches get into *my* house?"

"Roaches rely on people to survive. I am referring to the domestic roaches that you find in your home. They need food, shelter, warmth and high humidity, all of which you are providing. There are more roaches in America today than last year, and there will be more roaches next year. That is because we build more buildings and there are more places for roaches to live."

Sojourning cockroach

"But how did the roaches get *into* my house?"

"If you live in a city apartment, the chances are your roaches are the German kind. If they are, you probably carried them into your apartment yourself in your shopping bags and the cardboard boxes used to carry your food or furniture. It is also possible that while you were sitting on a bus or subway, a roach may have crawled into the cuff of your trousers and got a free trip home. If you work in a factory or your job requires you to hang your clothes in a locker next to those of other employees, it is possible for the roaches to crawl from one locker to another, and you may bring them home in your lunch pail or jacket.

"Roaches also follow water pipes and will move from one apartment to another, particularly if the people upstairs or downstairs moved and left no food. If a nearby apartment is being painted or treated for cockroaches, you can expect some migration in your direction."

It may be environmentally sound not to burn garbage in incinerators, but the alternative is compactors, which serve as excellent breeding grounds for cockroaches. The roaches then scoot from floor to floor, move out into the hallways and invite themselves into your apartment.

Pick up a six-pack.

Traveling on the cuff

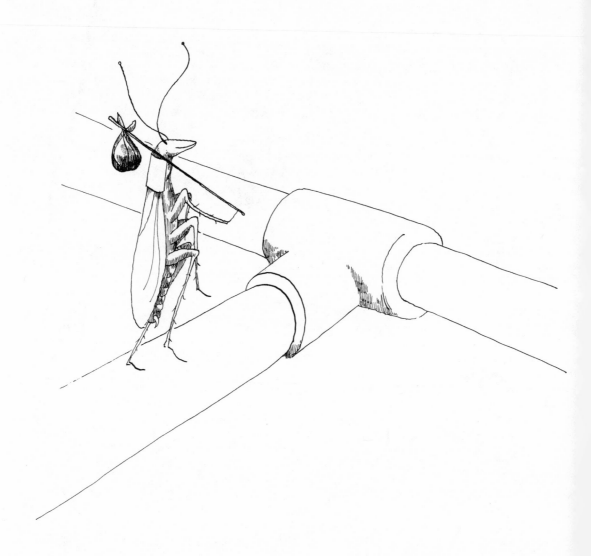

"Two pipes diverged upon the narrow wood, and I . . ."

Get them where they live.

8. DE-ROACHING MY HOUSE

I have roaches in my house. What should I do?

Answer: Try to control the roaches yourself by using certain pesticides that are sold over the counter. If you find this method doesn't work, call in a professional exterminator.

How do I find a qualified exterminator?

Answer: In the Yellow Pages of your telephone directory, you will find listings under "Exterminating and Fumigating (Pest Control)."

Once you've found the listing, you need to know if the exterminator is qualified. Reputable pest-control operators belong to certain professional associations. One is the 2,000-member National Pest Control Association, Inc. In order to join this organization, the exterminator must meet its standards of business ethics and pass a screening-committee review. There are also individual state and city associations in some large urban centers, such as the Greater Houston Pest Control Association. Members of these associations usually advertise in the local Yellow Pages and indicate in their ads that they are members.

However, if a pest-control operator does not belong to any of these associations, it doesn't necessarily mean he isn't qualified. If possible you should ask your pest-control operator if he is a member of any of these professional organizations; when in doubt you can telephone the association to verify the membership. I would also suggest finding out how long the operator has been in business. Three to five years is a minimum criterion.

How much do pest-control operators charge?

Answer: There are different ways that pest-control operators approach the fee they charge. They can give you a flat fee for the cleanout—$20 to $50 for a small apartment and sometimes as high as $90 + if the location is difficult to reach, is not on the operator's route, or they sell a Killmaster job guaranteed for a year. Sometimes a tenant may be moving on a certain day and his landlord, knowing that the apartment is heavily infested and having been pressured by other tenants, must treat the vacant apartment right away before the roaches have had time to scatter or the new tenants to move in. In such a case the operator may charge as high as $90 + because he must schedule the treatment at the tenant's discretion. A low fee of $15 may indicate competitive pricing and/or that the service has a technician in the immediate vicinity. The wide fluctuation in pricing more often has to do with the operator's lack of knowledge of his costs than with the quality of the service he gives.

Is there an advantage to signing a service contract?

Answer: In most cases, yes. If you have had to treat your home or apartment for cockroaches, you will probably get them again because of your location or your family's personal habits which allow them to enter. Under the service contract, each additional treatment may cost $7 to $15 plus the cost of the initial treatment. Within the first thirty days the pest-control operator will provide a free treatment as often as necessary if the homeowner or tenant complains about finding roaches after the initial treatment. Pest-control companies will recommend that you sign a service contract whereby they will treat your premises at regular intervals, usually monthly. In some areas of the country, especially the middle- and upper-income residential areas of the South, pest-control companies are starting to treat once every two months, spending more time at the job and giving a more thorough treatment. This decreases the companies' travel time and gasoline expenses. This is an efficiency move which benefits both the individual homeowner and the pest-control service.

In situations where there is a higher risk of having roaches, such as apartments which use compactors, most low-income areas and apartments located over supermarkets, taverns, bakeries, groceries, and other food-related establishments, treatment should be at least once a month, more often if needed.

Restaurants and food-processing plants should be treated at least once every two weeks and weekly as needed. This is because current pesticides will not remain efficient and continue to kill cockroaches in hot, food-oriented locations where the influx of new roaches is almost daily. There are longer-lasting pesticides, but these are not allowed to be used under state and federal regulations, as they present a health hazard.

What should I do before the pest-control operator comes to my home?

Answer: Roaches usually pocket in cracks and crevices near moist areas, particularly in the kitchen and bathroom. Therefore, it is essential that the pest-control operator treat all these areas. If he doesn't, you won't have a good job done or get rid of the cockroaches. If you have observed roaches in other areas, such as in bedroom closets, by the washing machine or in any other areas, be sure to inform the serviceman that cockroaches exist in these locations as well.

You must take out *all* pots, pans, dishes and silverware from all the cabinets. You can place these items in the living room, either on the floor or on a table—provided, of course, that there are no roaches in this area.

Then, take all the food out of the cabinets and place it in the living room too. As you are removing the food from the cabinet, you may find roaches inside paper receptacles (such as a napkin box, paper bags or food containers). Throw the containers into the garbage and remove the garbage from your premises immediately.

The next step is not essential but is recommended. Remove all excess spilled food (crumbs, etc.) from the cabinets, using a sponge or wiping cloth.

Then, go to the bathroom and empty out all items from the medicine cabinet and from under the sink. Be sure to remove any toothbrushes and toothpaste on the sink counter. Remove these items from the bathroom to allow the serviceman to apply the pesticide to as many areas as possible and as quickly as possible.

Subsequent treatment will usually not require that you remove these items again, unless you did not follow the directions properly the first time or you have failed to let the pest-control operator treat your home for several months.

What will the pest-control operator do?

Answer: He will apply a residual (long-lasting) pesticide onto surfaces, cracks and crevices so that wherever roaches crawl, they will come into contact with this chemical and die. Today's residual pesticides will be effective for approximately thirty days. Where cockroach populations are extremely high or difficult to control, the room will be treated with a fog to knock down the large numbers of cockroaches present and kill them. There are other space treatments which may also be employed. The serviceman may also use an irritant or flushing agent (such as pyrethrin) to search out cockroaches hiding in hard-to-reach locations. He may also supplement his sprays with dusts, tapes and/or baits.

If the kitchen and bathroom are prepared properly, treatment should take the technician between fifteen and thirty minutes. However, when computing the fee, you must include his preparation and traveling time.

How quickly can I put my dishes and food back?

Answer: Wait about ninety minutes or until you see that the treated surfaces are dry before replacing the items you removed from the kitchen and bathroom. Most important, *do not* wipe or wash off any of the dried chemical.

How do I know whether the treatment worked?

Answer: Over the next several days you can expect to see roaches crawling. However, these roaches will be picking up the pesticide and will die within several hours. If after two weeks you still see healthy cockroaches, then it is suggested that you request additional service.

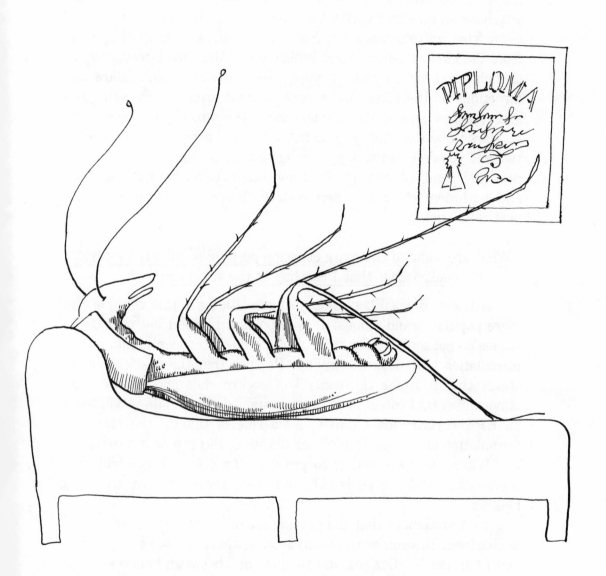

Understanding your cockroach

If I decided not to use a pest-control operator and wanted to treat my home myself, how would I go about it?

Answer: You need to think like a cockroach to control this creature effectively. There is still no easy solution. You must purchase an effective cockroach insecticide from a supermarket or even from an exterminator. You must still remove all the items from the kitchen cabinets and bathroom (as described previously in this chapter) before you can apply the pesticide. Then, follow the directions on the label. Make sure to treat areas where you have seen roaches besides the kitchen and bathroom. If your treatment appears to be working, you do not see cockroaches after a week or two and you don't want to go through all the bother of removing the pots and pans, silverware, food and other items from the kitchen and bathroom cabinets, then re-treat monthly according to the pesticide label.

What are some of the more common pesticides effective against cockroaches that I can buy at the supermarket?

Answer: Black Flag, Blackjack and Raid are just some of the more popular brand names. It is important to point out that these names do not indicate the type of pesticide, its concentration or the formulation that does the killing. Be sure that the pesticide label states that it is for the control of cockroaches and meets U.S. Environmental Protection Agency standards. Such labels will have an EPA registration number. As a rule of thumb, the chemical formulation should contain either diazinon, Baygon or dursban.

When you are looking to purchase an effective pesticide for cockroaches and are reviewing the label, there are two items to look for:

1. A statement that the percentage of active ingredients (such as diazinon, Baygon or dursban) is somewhere between ½ percent and 1 percent for diazinon and Baygon, and between ¼ percent and ½ percent for dursban.

ACTIVE INGREDIENT	IDEAL CONCENTRATION OF THE ACTIVE INGREDIENT
diazinon	0.5% to 1.0%
Baygon	same as above
dursban	0.25% to 0.5%

110

2. When purchasing an aerosol type of pesticide, look for a petroleum distillate content of less than 16 percent, with as low a content as possible. If you use aerosols with more than a 16 percent concentrate, you are playing with a flamethrower which might blow up in your hands if it comes into contact with your pilot light, cigarette or any other open flame.

What are pesticide bombs and how do they work?

Answer: The term usually refers to a one-shot treatment whereby the full contents of an aerosol container are released into an open space. The small pesticide droplets float into the spaces where cockroaches hide and kill them on contact.

Read the instructions on the label carefully because you will have to take additional precautionary measures, including leaving your home for at least several hours, turning off air-conditioners and fans, sealing air vents and other openings, and removing pets and fishtanks. On large fishtanks pull the plug of the aerator and seal up the top of the tank. About an hour after the room has been aired out, the aerator for the fishtank should be replugged.

Should I use roach traps such as Roach Motel?

Answer: These commercial devices will catch cockroaches, but will not eliminate the entire population and also appear not to be effective against the small, immature cockroaches. However, these traps are often helpful to use near pet cages (of gerbils, snakes, hamsters, birds, rabbits, etc.) where it is inadvisable to expose the animals to pesticides. These devices can often prove valuable indicators to ascertain whether a certain area of the home is infested with roaches.

Can I ever completely rid myself of cockroaches?

Answer: In a private house you have a better chance, but even here, if you continue to shop in the same grocery, work in the same factory and come in contact with more cockroaches, you can always bring new ones home. In apartment houses the chance of new cockroaches invading is even greater.

Completely ridding your home of cockroaches forever is not practicable, but existing populations in a specific house or apartment *can* be totally eliminated.

Some pest-control firms are now giving a six-month to one-year guarantee for one treatment. They say they will apply a pesticide with a paint brush. What are they using?

Answer: There is a new product out called Killmaster which contains 1 percent or 2 percent dursban in lacquer. It is relatively recent but seems to be effective. However, in apartments and commercial buildings where new cockroaches are continuously being brought in, more treatments are advisable.

Is boric acid the only safe thing to use around children?

Answer: Even boric acid is toxic when ingested in sufficient quantities. Treat it like any other pesticide. When boric acid is used as an eyewash, it is prepared as a 2.2 percent isotonic solution.

Are borax and boric acid the same thing?

Answer: No. Boric acid is derived from borax via a chemical reaction. Although borax exhibits some pesticidal activity, it is even slower acting than boric acid.

Will the use of boric acid completely rid a premise of cockroaches forever?

Answer: If we had such a marvelous item, the future of the cockroach would indeed be bleak. There are just too many areas (cracks, crevices, wet areas) where the cockroach can survive that are not possible to treat with boric acid.

Working the sandwich counter

113

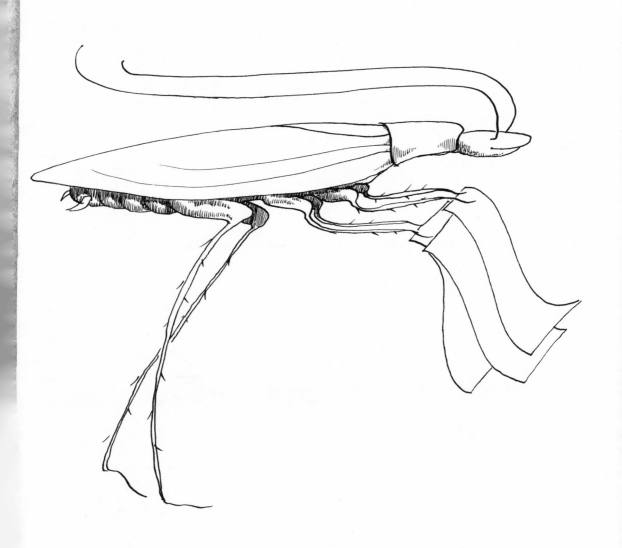

Masochistic cockroach presenting useful charts

9. USEFUL CHARTS

THE BATHROOM

Hot Spots to Check

- Medicine cabinets
- Pipes leading into walls
- Cracks near bathtub shower
- Behind radiator stand
- Stored paper goods
- Behind peeling wallpaper
- Hinges on door
- Near all moisture sources: sink, shower, toilet bowl
- Under bathroom scale
- Behind posters and notices
- Pleats on curtains
- Light switches
- In electrical appliances
- Corners of cabinets

Preparing for Treatment

- Remove pets and people.
- Empty medicine chest and remove items from room.
- Empty cabinets and remove all items from room.
- Don't forget to remove toothbrushes and personal belongings.

Treatment

- If you are treating area yourself, apply pesticide to all hot spots. Do not apply pesticide to areas where water or steam can come in contact.
- Professional exterminators may use a variety of approaches.
- Allow pesticides to dry for at least one hour before putting materials back.

Follow-Up

- Inspect hot spots daily for any signs of cockroach activity.
- Check all paper items brought into bathroom.
- Do not store unnecessary paper items in bathroom.
- Optional: Place two or three roach traps in areas where cockroaches were found.

THE KITCHEN

Hot Spots to Check

- Crack behind top of sink
- Pipes leading into walls
- Electrical appliances (examples: can opener, blender)
- Insulation of oven
- Motor casing of refrigerator and under rubber liner inside of door
- Cabinets storing dishes
- Wooden shelves
- Behind calendar and other paper items on wall
- Light switches
- In old newspapers, paper bags
- Garbage area
- Underside of chairs and tables
- Live plants
- Kitty litter
- Near animal cages
- Pet food

Preparing for Treatment

- Remove pets and people.
- Empty *all* cabinets, drawers and remove from room.
- Cover any pots that contain food.

Treatment

- Do *not* treat wooden chopping blocks and bread drawers with pesticides.
- Apply pesticides to all hot spots.
- Avoid water-base pesticides near electrical appliances.
- Avoid oil-base pesticides near pilot lights.
- Allow pesticides to dry for at least one hour before putting materials back.
- Use only registered pesticides and follow directions on label.

Follow-Up

- Inspect hot spots daily for any signs of cockroach activity.
- Check all paper items brought into kitchen.

- Do not store extra bags and newspapers in kitchen.
- Additional treatments may be necessary, so remove all items from shelves and drawers.

BEDROOMS AND LIVING ROOMS

Hot Spots to Check

- Around door frames
- Door hinges
- Around and in living plant material
- In wooden cabinets and shelves
- Inside TV and other electrical appliances
- Where food is stored
- In seams of chairs and sofa where food can accumulate
- Wet bar
- In bookshelves
- Behind bulletin boards
- Live plants
- Behind mirrors
- In cardboard boxes in closets or on shelves
- Light switches
- Pleats on curtains

Preparing for Treatment

- Remove pets and people.
- Unlock cabinets.
- Check hot spots and determine where moist areas are.
- Pull items off shelves and out of cabinets if infestation is bad.

Treatment

- If you are treating area yourself, apply pesticides to all hot spots as directed on the pesticide label.
- Professional exterminators may use a variety of approaches.
- Allow pesticides to dry for at least ninety minutes before putting materials back.

Follow-Up

- Inspect hot spots daily for any signs of cockroach activity.
- Optional: Place two or three roach traps in areas where cockroaches were found.

OFFICE BUILDINGS
(Professionals, dentists, lawyers, bankers, insurance agents, brokers, etc.)

Hot Spots to Check

- Water cooler
- Coffee area
- Cafeteria
- Bathrooms
- Office refrigerator
- Vending machines
- Mop closet/slop sinks
- Locker area
- Live plants
- Electrical appliances
- Near fishtanks
- Nurses' quarters
- Mail room

Preparing for Treatment

- Thoroughly inspect hot spots.
- Determine which areas need treatment.
- Where necessary, remove people. In some cases this is not necessary.
- Contact a professional pest-control company.

Treatment

- Will vary greatly with the degree of infestation and areas needed to be treated.
- Advise use of odorless chemicals in areas where people work.
- Assist serviceman by telling him exactly where and when you saw a problem.

Follow-Up

- See that report pad is maintained. Alert outside service technicians in order to locate new problems easily.
- Optional: Place roach traps in inspected areas to monitor results.

RESTAURANTS, SUPERMARKETS, CAFETERIAS AND OTHER FOOD-PROCESSING FACILITIES

(Part 1: Food-Processing Areas)

Hot Spots to Check

- Dishwashing area
- Ice machines
- Inside all food-processing machinery (ovens, refrigerators, etc.)
- Under and behind electrical appliances (clocks, radios)
- Food carts
- Wet-bar area
- Steam tables
- Sinks
- Suspended ceiling
- Serving stations
- Incoming materials
- Space does not permit a longer list

Preparing for Treatment

- Will vary with type of treatment used.
- Sometimes requires at least two hours' shutdown time.
- Unlock all areas that need treatment.
- Record as specifically as possible all areas where cockroaches were found.

Treatment

- Requires the services of a certified professional pest-control technician.

Follow-Up

• Communicate with your pest-control technician.

RESTAURANTS, SUPERMARKETS, CAFETERIAS AND OTHER FOOD-PROCESSING FACILITIES

(Part 2: Non-Food-Processing Areas)

Hot Spots to Check

• Locked closets
• Storage shelves
• Rest rooms
• Cash register
• Water cooler
• Electrical boxes
• Boxes in storage
• Slop sinks
• Basement area
• Space does not permit a longer list

Preparing for Treatment

• Will vary with type of treatment used.
• Sometimes requires at least two hours' shutdown time.
• Unlock all areas that need treatment.
• Record as specifically as possible all areas where cockroaches were found.

Treatment

• Requires the services of a certified professional pest-control technician.

Follow-Up

• Communicate with your pest-control technician.

Pet shops, zoos

Where to Inspect

- Top and bottom of cages along seams
- Behind walls housing cages
- Animal-feed storage areas
- All wet areas
- See chart on kitchens for more details

Professional Tips

- Feed animals during day. Do not leave food around at night for cockroaches to feed on. When working with professional pest-control operators, inform them of which animals are most expensive species present.

Precautions

- Where possible, avoid use of pesticides that have vapors. Keep baits away from animals, particularly monkeys.

Areas with Pets

Birds
- Particularly sensitive to space sprays; remove birds before treating.

Cats
- Avoid contact with any pesticide dust. Cats continually groom themselves and will ingest the poison.

Dogs
- The smaller the dog, the more sensitive to pesticides.

Reptiles
- More tolerant to organophosphates than most animals.

Animal research facility

Precautions

- May require *no* pesticides—sanitation and traps.

Hospitals (operating room, patient area)

Where to Inspect

- Hollows of beds
- Around sinks
- In ceiling, especially if ceiling is suspended
- Night stands
- Closets

Professional Tips

- Proper sanitation is very important.
- Patch cracks and crevices. Work closely with a professional pest-control operator.

Precautions

- Must sterilize when finished using any pesticide. Best to remove infested machinery to another area and treat somewhere else.

In hospitals and out-patient clinics

Precautions

- Avoid pyrethrin-base products, as they are particularly irritating to persons with respiratory problems.
- Avoid placing baits, tapes or other solid items where patients can reach them.

Items used for transportation (planes, boats, trains, buses)

Where to Inspect

- All food areas
- All rest rooms
- Storage facilities

Precautions

- Where possible, avoid treating when people are present.

Professional Tips

- Proper sanitation is very important.
- Patch cracks and crevices. Work closely with a professional pest-control operator.

Schools

Where to Inspect

- Cafeteria
- Home economics area
- Vending machine areas
- Gymnasium
- Classrooms
- Biology laboratories where live animals are kept

Professional Tips

- Proper sanitation is very important.
- Patch cracks and crevices. Work closely with a professional pest-control operator.

Precautions

- Treat when students are not present.
- See chart on commercial kitchens for restrictions on use of pesticides.

Apartment houses

Where to Inspect

- See charts on bedrooms, bathrooms and kitchens.
  Also:
- Incinerator-compactor room
- Compactor chute
- Slop sinks
- Basement area
- Laundry area
- Neighbor who will not let pest-control technician enter to treat properly

Professional Tips

- Temporary measure of stuffing steel wool around pipes does decrease cockroaches coming from adjoining apartment.
- Work with a professional pest-control operator.
- Give him access to all areas that need treatment.
- Do not leave garbage in hallways.

125

restroom

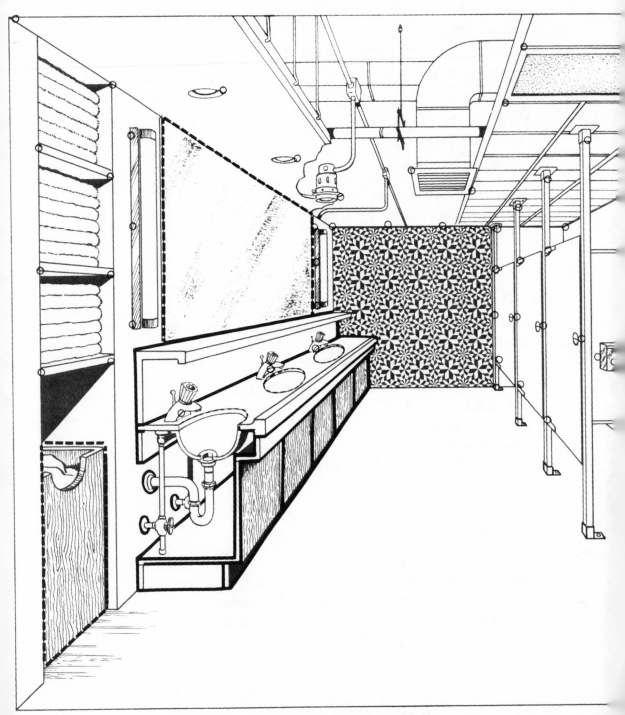

Sanitation Suggestions as Related to Pest Management

Empty and clean daily all refuse containers, soap dispensers, urinals, wash stands, toilets and mirrors on walls.

How Insects Enter

They come in on people's baggage, paper sacks, cartons of paper products, through cracks in floors and walls, on carts that transport supplies, through open windows and doors, and through floor drains. They breed in wastes in or behind lockers, along walls, floor drains and pipe traps.

Roach Location Chart

▬▬▬▬▬▬	1st most likely locations where roaches will be found
▪▬ ▬ ▬ ▬ ▪	2nd most likely locations where roaches will be found
○○○○○○	3rd most likely locations where roaches will be found

bookkeeping and office

Sanitation Suggestions as Related to Pest Management

Persons responsible should discard all food from desks, lockers, files, and other cabinets. Thoroughly clean up all food residue from the area. Don't keep food in this area.

How Insects Enter

Insects enter through open doors, windows; cracks around doors and service lines, cracks in walls, on food carried into the area, on plants, people's clothing, cartons, packages, service carts and equipment.

Roach Location Chart

▬▬▬▬▬ 1st most likely locations where roaches will be found

▬ ▬ ▬ ▬ ▬ 2nd most likely locations where roaches will be found

○○○○○○ 3rd most likely locations where roaches will be found

Roaches Require Food, Warmth Moisture, and Shelter

Roaches come from their hiding places to find food and moisture in order to survive. The Whitmire Guide points out the most common locations where roaches are found in hospitals and nursing homes. However, roaches may be found in any location where their survival requirements exist.

The label directions and caution statements as printed on each Whitmire Prescription Treatment product should be followed when treating hospitals and nursing homes.

129

coffee shop

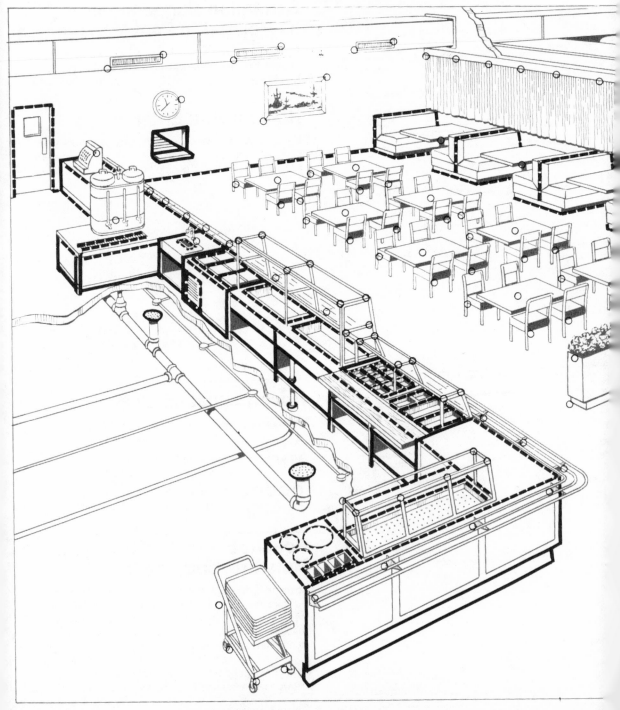

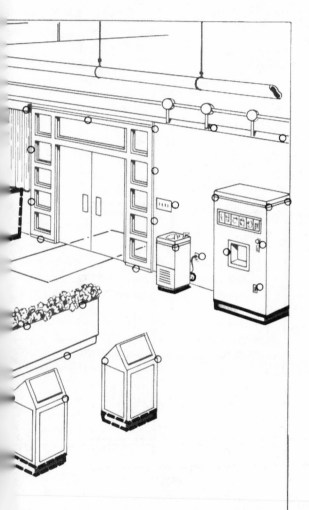

Sanitation Suggestions as Related to Pest Management

In this area insects can develop rapidly anywhere food and beverage wastes are allowed to accumulate. Strict sanitation is mandatory to prevent heavy insect population buildup. Empty and wash all waste containers daily. Seal all cracks and crevices. Daily clean food droppings from between and under all serving areas. Keep area dry and drains cleaned.

How Insects Enter

They follow service lines, such as drains, steam, water or electric, from areas of infestation and enter through the wall or floor where these service lines enter. Eggs or insects are carried in on cartons or baggage of workers or visitors.

Roach Location Chart

━━━━━ 1st most likely locations
where roaches will be found

▮ ▬ ▬ ▬ ▬ ▬ ▪ 2nd most likely locations
where roaches will be found

○○○○○○ 3rd most likely locations
where roaches will be found

coin machine
food service

Sanitation Suggestions as Related to Pest Management

Keep the inside of vending machines clean and repaired. Keep floor clean and dry under, between, and around vending machines. The heat, humidity, spilled food, and the many cracks for hiding make this an ideal area for roaches and a breeding area for flies. Beverage tank areas must be cleaned on a routine schedule. Cartoned beverages and supplies must be properly organized, inspected, and stock rotated.

How Insects Enter

Insects come in on service refill cartons; they follow the electric, water and heat service lines and come in through water drainage lines. They breed in hidden waste, work their way in through cracks from outside walls.

Roach Location Chart

———— 1st most likely locations where roaches will be found

- - - - - - - 2nd most likely locations where roaches will be found

OOOOOO 3rd most likely locations where roaches will be found

housekeeping

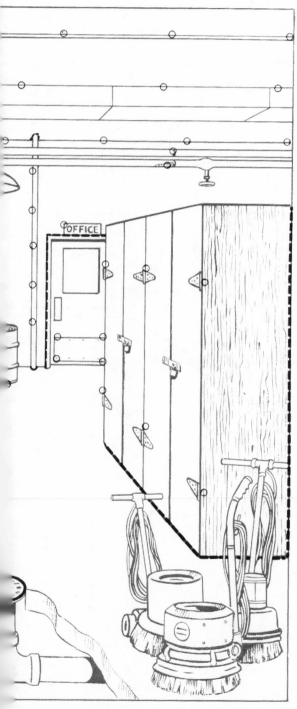

Sanitation Suggestions as Related to Pest Management

Keep supplies stacked neatly and stored off of floors and away from walls. Rotate inventory. Seal all cracks and crevices. Hang used mops and brooms. Empty into sealed bags all used cartons which may be carrying insects or their eggs. Keep area clean and dry at all times; keep foods out of this area. Do not transport empty supply boxes and cartons to any other areas within the complex.

How Insects Enter

Insects come into the area through open windows, doors, or through cracks in walls, ceilings, or floors. Insects or their eggs come in on cartons of material and on service carts. They breed in hidden waste.

Roach Location Chart

———————— 1st most likely locations where roaches will be found

– – – – – – – 2nd most likely locations where roaches will be found

○○○○○○ 3rd most likely locations where roaches will be found

locker room

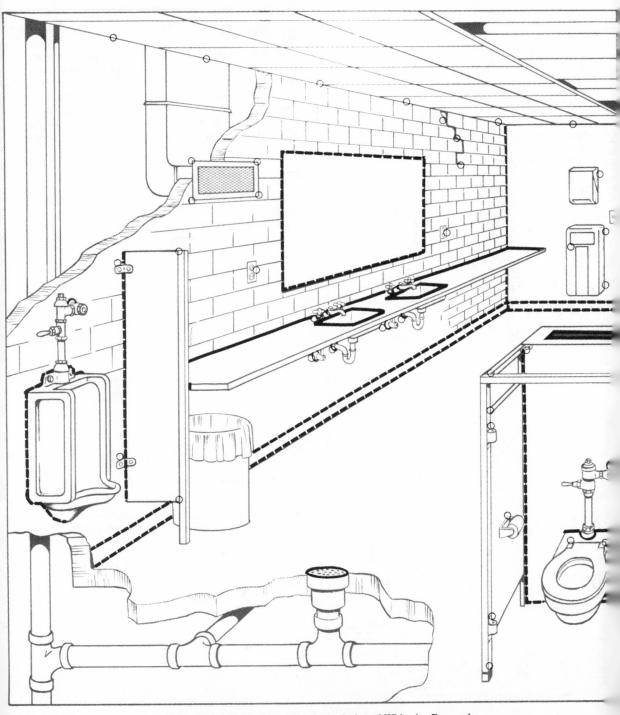

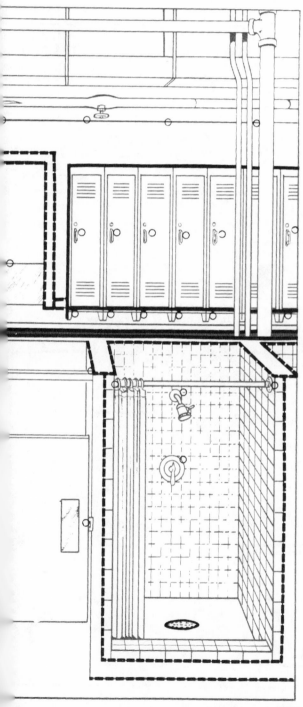

Sanitation Suggestions as Related to Pest Management

Empty and clean daily all refuse containers, soap dispensers, urinals, wash stands, toilets, showers and mirrors on walls. Empty and clean lockers regularly—keep all food and beverages out of lockers. Seal all cracks.

How Insects Enter

They come in on people's baggage, paper sacks, cartons of paper products, through cracks in floors and walls, on carts that transport supplies, through open windows and doors, and through floor drains. They breed in wastes in or behind lockers, along walls, floor drains and pipe traps.

Roach Location Chart

━━━━━━━ 1st most likely locations where roaches will be found

▬ ▬ ▬ ▬ ▬ 2nd most likely locations where roaches will be found

○○○○○○ 3rd most likely locations where roaches will be found

laundry

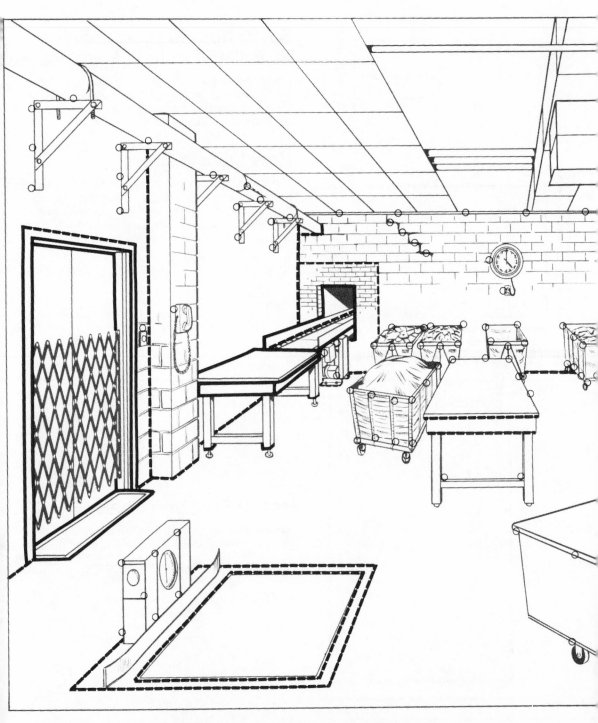

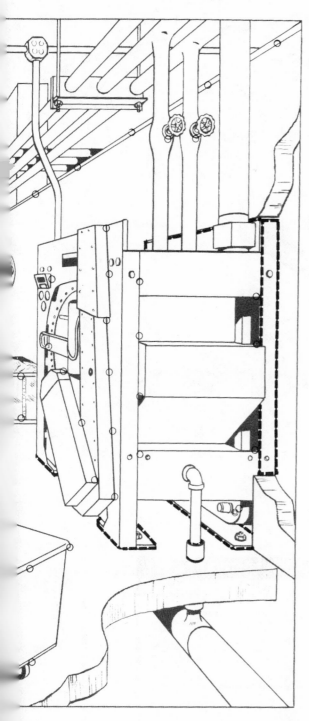

Sanitation Suggestions as Related to Pest Management

Keep floors clean; seal all cracks and crevices, and clean laundry carts regularly.

How Insects Enter

Insects enter on dirty laundry, through cracks in walls, through doors or elevator shafts, and breed on carts or in waste carried in on workers' baggage.

Roach Location Chart

——— 1st most likely locations where roaches will be found

- - - - - - 2nd most likely locations where roaches will be found

○○○○○○ 3rd most likely locations where roaches will be found

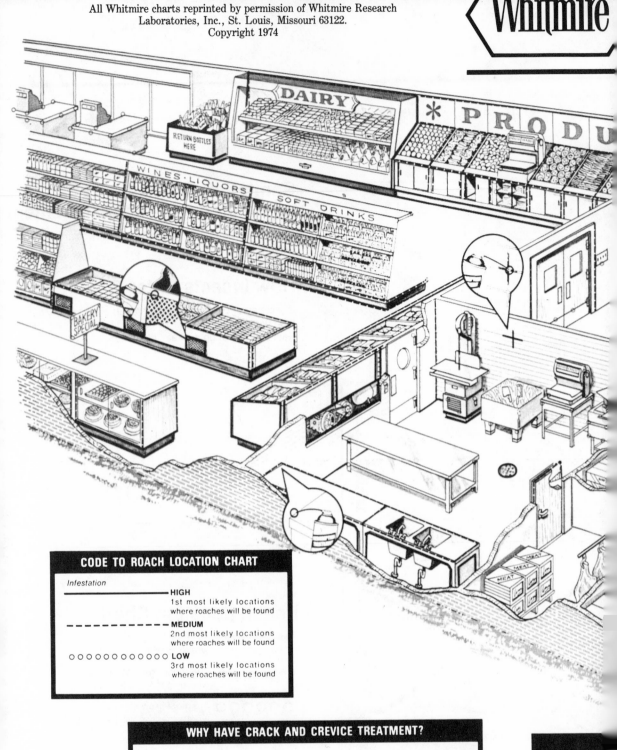

CODE TO ROACH LOCATION CHART

Infestation

—————— **HIGH**
1st most likely locations
where roaches will be found

- - - - - - - - - **MEDIUM**
2nd most likely locations
where roaches will be found

○○○○○○○○○○○○ **LOW**
3rd most likely locations
where roaches will be found

WHY HAVE CRACK AND CREVICE TREATMENT?

1. It places small doses of insecticide directly into areas where roaches live, breed, and are most likely to be found.

2. Placing the insecticide into the cracks and crevices produces roach control with reduced environmental contamination.

3. Crack and crevice treatment places the insecticide beyond the reach of children and pets.

4. Small amounts of insecticide placed into cracks and crevices reduces odor and staining problems.

5. The placement of chemicals into cracks and crevices protects them from mechanical removal such as vacuuming and mopping; thereby, extends their life.

6. Chemicals placed into cracks and crevices are protected from light and air movement; thereby, extending the life of the chemicals by reducing chemical oxidation and evaporation.

7. The professional way of roach control is better placement of the pesticide chemicals being used.

The stateme

APPLICATI

Limited to C
directly into
ments of co
housing, jur
roaches hid
surfaces or
food proces

Applications
other than

PRESCRIPTION TREATMENT™ System
Crack and Crevice Injection Method for Roach Control®

ROACH LOCATION CHART
for Super Markets

ROACHES REQUIRE FOOD AND WATER

Roaches come from their hiding places to find food and moisture in order to survive. The Whitmire Guide points out the common locations where roaches are found in Supermarkets. Roaches may be found in any location where their survival requirements exist.

SPECIAL NOTICE:

Sections of Supermarkets are food preparation areas or edible product areas of food plants. Read the directions and warning statements printed on each Prescription Treatment product label before using and use only those products labeled for such in those locations.

PORTANT NOTICE

the following Prescription Treatment products:

atment No. 250 — 1.0°₀ Baygon
atment No. 260 — 1.0°₀ Diazinon
atment No. 270 — 0.5°₀ Dursban

AREAS OF FOOD HANDLING ESTABLISHMENTS

reatment Only — Apply a small amount of material
s such as expansion joints between different ele-
n equipment bases and the floor, wall voids, motor
boxes, conduits or hollow equipment legs where
ken to avoid depositing the product onto exposed
rials into the air. Avoid contamination of food or

the food areas of food handling establishments,
e treatment, are not permitted.

Form No. 02 55 Copyright 1974, WHITMIRE RESEARCH LABORATORIES, INC., St. Louis, Missouri 63122

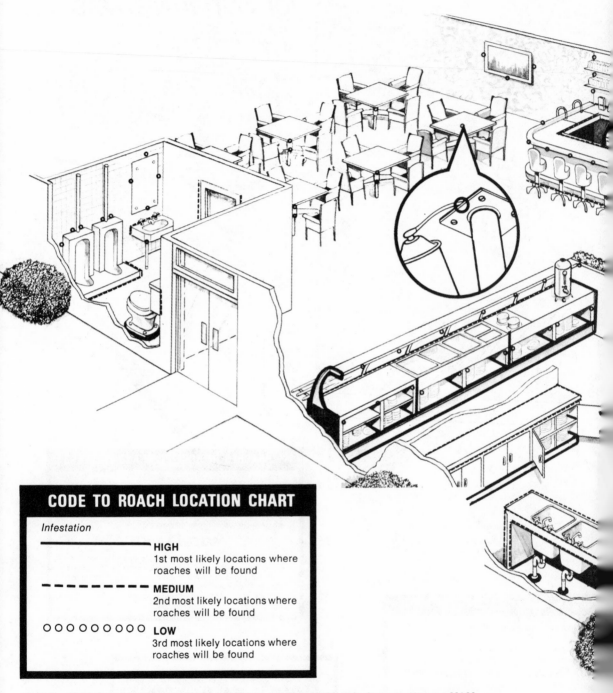

CODE TO ROACH LOCATION CHART

Infestation

HIGH
1st most likely locations where
roaches will be found

MEDIUM
2nd most likely locations where
roaches will be found

○○○○○○○○○ **LOW**
3rd most likely locations where
roaches will be found

Form No. 02-55 Copyright 1974, WHITMIRE RESEARCH LABORATORIES, INC., St. Louis, Missouri 63122

PRINTED IN U.S.A.

ROACH LOCATION CHART
for Restaurants · Lounges · Bars

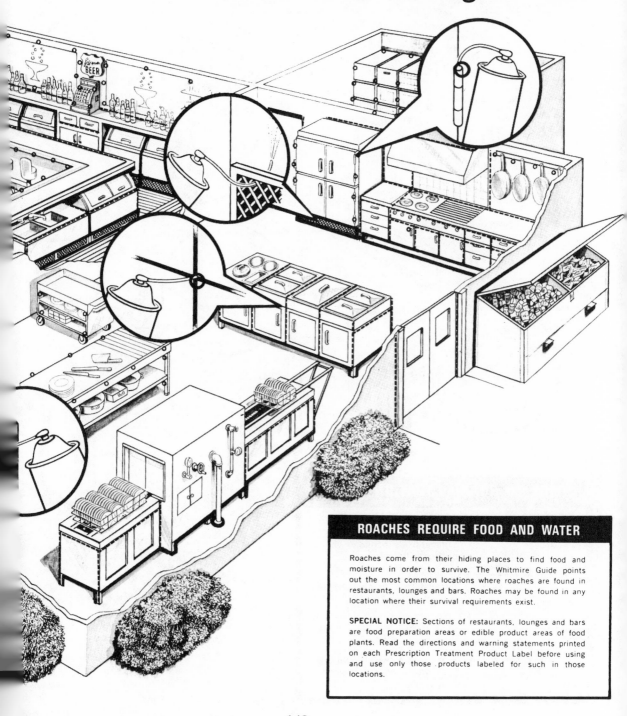

ROACHES REQUIRE FOOD AND WATER

Roaches come from their hiding places to find food and moisture in order to survive. The Whitmire Guide points out the most common locations where roaches are found in restaurants, lounges and bars. Roaches may be found in any location where their survival requirements exist.

SPECIAL NOTICE: Sections of restaurants, lounges and bars are food preparation areas or edible product areas of food plants. Read the directions and warning statements printed on each Prescription Treatment Product Label before using and use only those products labeled for such in those locations.

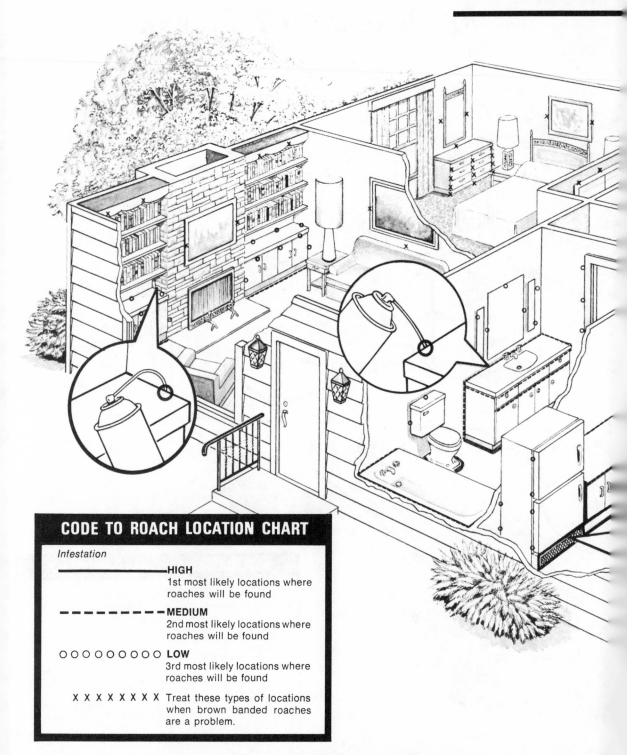

CODE TO ROACH LOCATION CHART

Infestation

━━━━━━━━━━━━━**HIGH**
1st most likely locations where
roaches will be found

━ ━ ━ ━ ━ ━ ━**MEDIUM**
2nd most likely locations where
roaches will be found

○○○○○○○○○ **LOW**
3rd most likely locations where
roaches will be found

X X X X X X X X Treat these types of locations
when brown banded roaches
are a problem.

PRESCRIPTION TREATMENT™ System

Crack and Crevice Injection Method for Roach Control®

ROACH LOCATION CHART
for Homes and Apartments

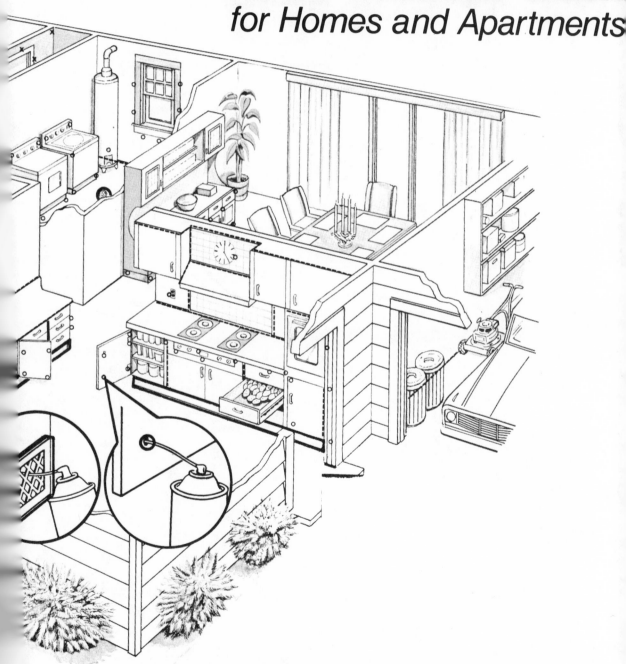

145

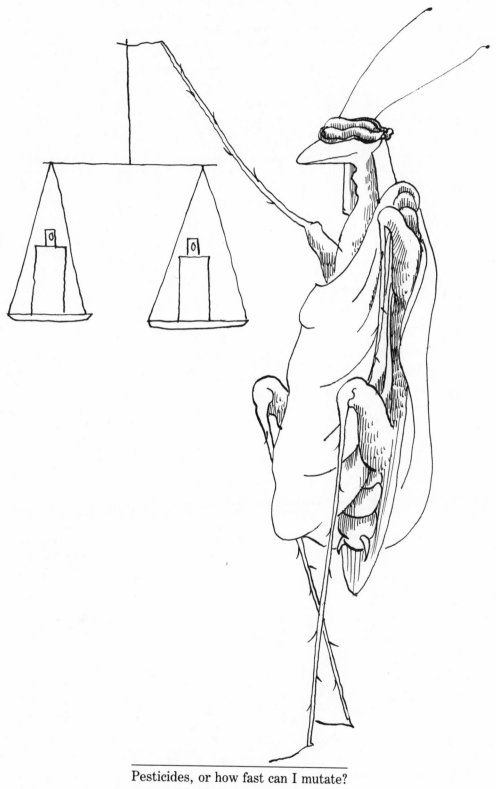

Pesticides, or how fast can I mutate?

Aerosols

The most common types of aerosol familiar to the consumer are Raid, Ortho, Black Flag and Black Jack cockroach sprays. In certain areas one will find similar ingredients packaged under local trade names. When buying aerosol sprays the consumer should consider the following:

1. the label on the pesticide container shows it also controls cockroaches;

2. the label should state that it can be used for cockroaches indoors; and

3. the higher the percentage of active ingredients shown on the label, the greater the kill. For example, a 1 percent concentrate of the active ingredient will be ten times more potent than 1/10 of 1 percent, or a 0.1 percent concentrate.

The most widely used active ingredients are:

1. *Pyrethrin*, a quick knockdown or irritant agent that flushes the roaches out but usually does not kill them.

2. *Diazinon* is a long-lasting (residual) spray that will continue to kill for up to thirty days. The consumer should be careful to note that diazinon and other active ingredients in the aerosol spray *may* be listed under their chemical or trade names, and for this reason we have prepared the following chart to show the common (generic), chemical and trade names.

3. *Baygon* (often spelled BAYGON because it is an accepted trade name) is a long-lasting (residual) spray that will continue to

kill up to thirty days. Baygon also acts as an irritant and flushes out roaches better than any other residual agent.

4. *Dursban* is probably the longest-lasting residual spray used under normal conditions.

Full or total-release bombs

Full or total-release bombs are a special type of aerosol in which all the pesticide is released at one time. It is necessary to leave the room immediately after releasing the bomb. For best results the premises should be kept tightly closed for two hours. All windows, air vents and holes should be closed. People and pets must not be present at any time during this operation. These bombs contain VAPONA or pyrethrin. Both of these ingredients are knockdown agents. There are a few bombs available which also contain a residual pesticide. Some of these bombs are available only to the professional exterminator.

Before using this technique one should open up all cabinets, drawers and other storage areas where cockroaches are suspected of hiding. Be sure that the bomb is at room temperature (65° or more F.) before using. If the temperature of the ingredients is too low, the contents will sputter or be emitted as large droplets. Place a blank piece of paper on top of a chair or in a high area, and then put the bomb on top of the paper. The paper will serve to protect the surface adjacent to the bomb. Do not use newspaper because the print may stick to the furniture.

Dangers of aerosols

If the petroleum distillate content of the container is higher than 12 percent to 15 percent, we are dealing with a highly dangerous package that can become a flamethrower. Also do not smoke while using an aerosol and stay away from pilot lights or other flames.

COMMON NAME	CHEMICAL NAME**	SOME TRADE NAMES** OR OTHER DES-IGNATIONS	R.	C.
bendiocarb*	2,2-dimethyl-1, 3-benzodioxol-4-61 methylcarbamate; an alternative name is 2,3 isopropyli-dene dioxyphenyl methylcarbamate	Ficam	X	
propoxur	0-isopropoxyphenyl methylcarbamate	Baygon	X	
chlorpyrifos	0,0-diethyl 0-(3,5,6-trichloro-2-pyridyl) phosphorothioate	Dursban	X	
diazinon	0,0-diethyl 0-(2-iso-propyl-6-methyl-4-pyrimidinyl) phos-phorothioate	Diazinon, Spec-tracide	X	
pyrethrins	pyrethrin I and II	Numerous products con-tain pyrethrin		X
resmethrin	(5-benzyl-3-furyl) methy cis, trans-(±)-2,2-di-methyl-3- (2-meth-ylpropenyl) cyclo-propane-carboxyl-ate	SBP-1382		X
dichlorvos	2,2-dichlorovinyl di-methyl phosphate	Vapona, DDVP		X

R. = Residual   C. = Contact

*Not available for public use. Used only under the direction of certified pesticide applicators.

**The public will find the chemical name and the trade name on the pesticides they purchase.

Why sprays don't work

Liquid sprays or emulsions

The homeowner will rarely use liquid emulsions (liquid sprays) for two reasons. First, the liquid concentrate is very toxic and it would be dangerous to have such a concentrate around the house where a child could come in contact with it and swallow it. Second, a special piece of applicator equipment must be used, and this is costly for a homeowner to use only once or twice. This piece of equipment is called the compressed handsprayer.

The professional who may use the liquid sprays will refer to them by their commercial names—Straban, Diazinon 4E or Dursban 2E. Because of the danger of liquid-spray concentrates, the homeowner should steer clear of mixing his own concentrates and leave their use to the professional.

Dusts

The consumer is familiar with using dusts such as boric acid to combat cockroaches. A recent and well-popularized commercial boric acid is Roach Prufe. Dusts are generally used by the professional pest-control operator to supplement liquid sprays. Some of the more popular dusts are pyrethrin, diazinon, silica gel and boric acid. The professional relies more heavily on the first because of its lack of odor, and the fact that roach populations have not developed a resistance to this chemical despite the numerous years that it has been in use.

Boric acid, when used as a dust, does not float into the atmosphere as much as does silica gel. This minimizes its inhalation by humans and the contamination of the premises. However, when blown back into wall cracks, crevices and voids, it does not give as good a coverage as silica gel.

There is no perfect pesticide. The disadvantages of boric acid are as follows: (1) it's unsightly; (2) if it becomes wet it is ineffective and since most cockroaches congregate in moist areas, this limits its use in destroying the population; and (3) it is slow acting and will often take from four to seven days to kill the roach. The kill is accomplished by the roaches' picking up the dust on their antennae and feet and then ingesting it (by preening their bodies). Death can occur in children and adults by accidentally ingesting considerable quantities of the dust.

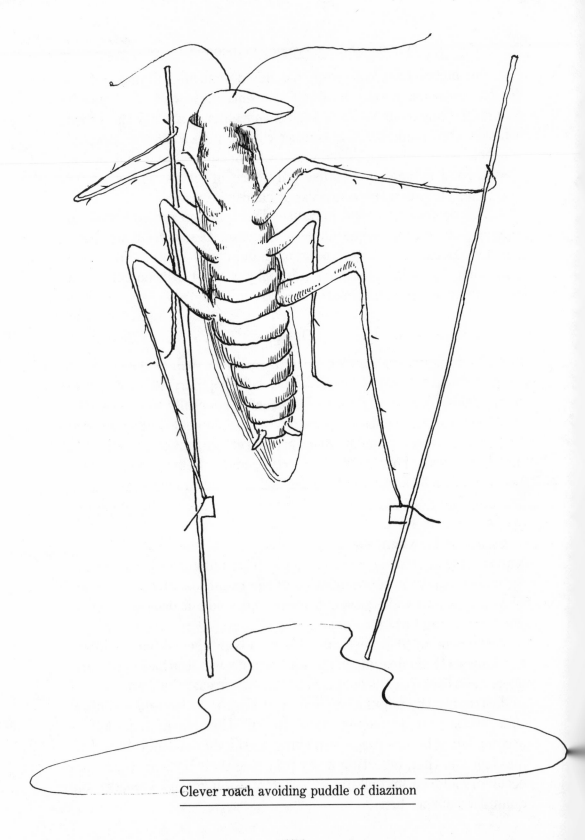

Clever roach avoiding puddle of diazinon

Baits

The most popular and probably the best bait is Bolt Roach Bait (which contains dursban). Baits are edible materials used to lure the roaches to a given area. They ingest this material and die because of its toxicity. Baits are relatively recent on the market, and new ones are being continually developed. One of the most effective baits, Kepone, was taken off the market.

Baits work best on the larger cockroaches. Although it takes longer to kill cockroaches with baits than with sprays or aerosols, practical experience shows us that when baits accompany the use of sprays, one can expect better control than by using either method alone. Baits are particularly effective where there is a lack of readily available food for the cockroaches (for example, in clothes closets).

The most easily obtainable bait is Bolt Roach Bait. This bait contains 0.50 percent dursban. Professional pest-control operators use 2 percent Baygon bait. In either event the key to effective use is to apply the bait in many locations and keep it dry.

Mechanical traps

Mechanical roach traps (made of metal or paper) go under a number of trade and commercial names such as Roach Motel, Roach-Trap and Mr. Sticky.

All traps rely on the same principle to work. They count on the fact that roaches like to crawl into cracks and crevices. The construction of the trap entices the roach to enter, and once inside, they are trapped by a glue-like substance.

However, we cannot evaluate the success of a roach-control program by looking at how many are killed but rather at how many are left alive. This is because roaches breed at a prolific rate; for instance, the German cockroach breeds at the rate of forty young or more every twenty-eight days.

In most cases traps do not eliminate cockroach populations, but do serve two useful purposes: (1) They reduce numbers, and (2) they let one know in what areas cockroaches still pocket. The paper traps are cheaper and much more effective than the metal ones.

155

Very fast roach avoiding pyrethrin bomb

Roach tape

One of the newest methods for controlling cockroaches is the slow-release roach tape. The tape contains a pesticide, and pieces of it are placed in hard-to-reach areas. Hopefully, the roaches will walk over the tape and die. The uses for this type of tape are still being explored but have shown some promise.

ULV

ULV stands for "Ultra Low Volume." It is also commonly referred to as ULD ("Ultra Low Dosage").

This method takes concentrated pesticide and breaks it up mechanically into minute particles ideal for killing cockroaches. These small particles float through the air until they reach and kill the cockroaches. Only professionals should use this method; it is not available to the public. The professional must use a respirator with this method, and the premises must be sealed and evacuated for at least two hours.

The piece of equipment used for ULV treatment is the aerosol generator. This should not be confused with the pressurized aerosol cans that the public can buy at the supermarket.

Fogging and misting

There are mechanical and thermal machines which apply pesticides so that the droplets emitted disperse as a mist or fog. This method should be used only by professionals.

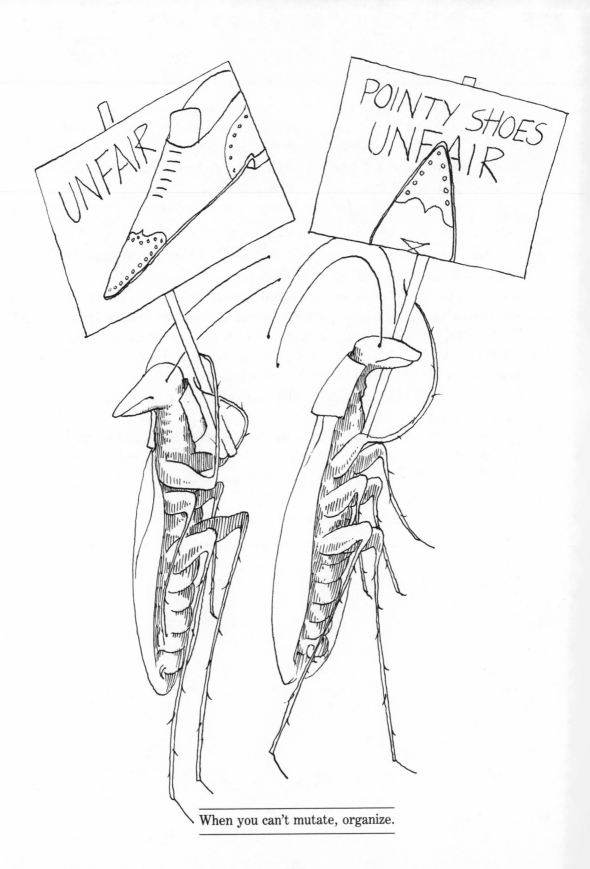

When you can't mutate, organize.

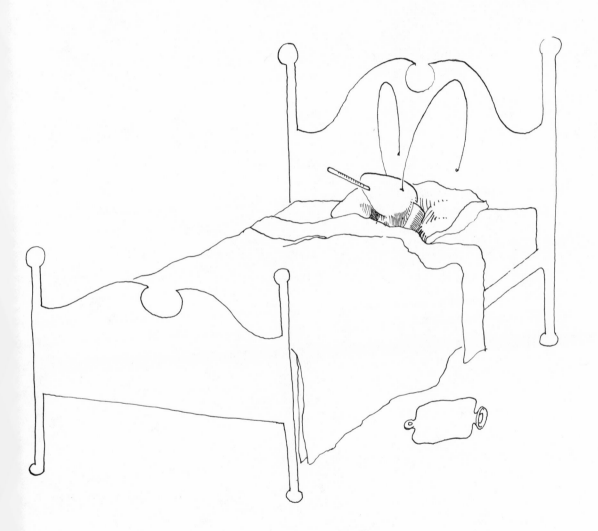

How to stay healthy

Inhaling pesticides

If the pesticide contains pyrethrin, a common ingredient in many household pesticides, it can cause wheezing, tearing of the eyes, throat irritation and coughing. Those suffering from asthma and other respiratory conditions are affected most. The pesticides which the U.S. Environmental Protection Agency (EPA) allows to be registered for the control of cockroaches indoors must be formulated so that slight inhalation will not be deadly. To date, all studies on pyrethrin and other pesticides sold for cockroach control confirm that there is no long-term accumulation of the pesticide in the human body, and that our bodies break down and detoxify any small amount of pesticide we might inhale.

Tips on keeping pesticides away from children and pets

People often ask, "How do I keep pesticides away from children and pets?" First, buy only what you need and try to use the pesticide the same day you buy it or as soon as possible. Second, if you have to store pesticides, keep them out of the reach of children and pets. If at all possible, keep the chemicals in a locked area. When you use a pesticide, remove pets and children from the room or area so that they do not see where you place the pesticide.

Pesticide odors

It is often the impurities within the pesticide formulation which make it smell so bad. The cost of purifying the formulation would make the price prohibitive. Recently, manufacturers have developed new pesticides which are odorless, but these formulations are unavailable except to professional pest-control operators.

Pesticidal effects on pets

All pesticides are poisonous. However, the formulation is aimed at killing cockroaches, which weigh much less than your household pets. If you were to leave the insecticide around so that pets were to ingest or inhale large amounts, it could make them ill. Tropical fish and birds will be more sensitive than cats or dogs. I have seen fish die from cockroach sprays because the pesticide applier forgot to turn off the air pump, which then sucked the pesticide vapors directly into the fishtank.

Accidental contamination

What would happen if some pesticide spray got on your toothbrush? The first time you brush your teeth it might not taste so good, but thereafter you should expect a normal taste if you have washed your toothbrush off. One should try to avoid spraying food, toothbrushes, plates and other such items. All pesticides are poisonous. Because it is going to be used around pets, plants and people, it is necessary to restrict the use of what kind of pesticide can be used in the home. Keep in mind that the pesticide has been formulated to kill an animal that weighs less than your little finger. Therefore, the amount which you have sprayed on your toothbrush is, hopefully, not enough to do any damage. Furthermore, your body will break down the minute amounts of pesticide you may ingest. However, if you buy the *wrong* kind (one not designed to be used indoors), you can become quite ill and you should consult a doctor. The less you weigh, the more the pesticide will affect you because pesticide kill is based upon body weight of the victim.

Behavior of cockroaches affected by pesticides

Cockroaches have a nervous system that deteriorates when exposed to an insecticide. Once the roach loses its sense of balance, it falls over on its back, becomes disoriented and cannot right itself.

There are some commercial insecticides, particularly pyrethrin, which on contact with the cockroach cause temporary

162

paralysis. However, the concentration of the chemical is insufficient to kill, and when the cockroach recovers he scurries away. It appears that he has "played possum."

To control roaches properly we want to use a long-lasting residual pesticide that continues to work for up to a month. These pesticides do not kill the roach on contact but rather affect its nervous system. After the roach comes in contact with the pesticide, it can take more than an hour before it begins to show ill effects and several hours will pass before the roach dies. Those pesticides which are supposed to kill roaches on contact are more expensive and stop working after a few hours, and in many cases, the so-called "dead roach" gets up and walks away.

Cockroaches and disease

Recent research has shown that cockroaches have the greatest potential of all domestic insects for transporting bacteria that cause human food poisoning. Cockroaches in clean apartments carry just as many bacteria as those in filthy apartments. Some of the bacteria carried by cockroaches are *Streptococcus, coliform, Salmonella, Staphylococcus* and *Clostridium.*

Recent research indicates that small nymphs of the German cockroach are capable of penetrating sterile packs stored on shelves prepared for immediate surgery. To complicate the matter, they are easily capable of transporting bacteria to the sterile utensils painstakingly prepared for surgery.

If cockroach populations are allowed to go unchecked, tremendous numbers occur. Because of lack of food, cockroaches may begin to feed on the eyebrows, sores and nails of young children and bedridden adults. Feeding is done at night while the person is asleep.

Cockroaches possess oils which cause some persons to break out in hives. Research has shown that boiling food which the cockroaches have walked over does not destroy the allergy-producing substance.

Meet the exterminator

The leading professional associations are listed below. Their criteria for membership involve business-ethic codes, certification requirements, number of years in business and legal and insurance standards. Although many qualified exterminators may not be members of these organizations, it is nevertheless recommended that one do business with established and certified companies.

There is one national association of pest-control operators which includes a professional staff of technically trained entomologists. Each week they forward to their members bulletins covering the latest developments in the field, legislation, equipment and new products. This organization is:

THE NATIONAL PEST CONTROL ASSOCIATION, INC.
8150 Leesburg Pike, Vienna, Virginia   22180 (703) 790-8300

NPCA MEMBERS—NORTH AMERICA & FOREIGN
Directory—Who's Who—By Member Firm, National Pest Control Association. Look for their logo in the Yellow Pages.

ARIZONA
Pest Control Operators of Arizona; Ms. June Weathers, Secretary-Treasurer; 6657 Calle Dened; Tucson, Arizona 85710.

CALIFORNIA
Pest Control Operators of California; Ms. Gussie Hayes; 3444 West First Street; Los Angeles, California 90004

HAWAII

Hawaiian Pest Control Association; Ms. Marjorie Carter, Secretary; 2140 Kaliawa Street; Honolulu, Hawaii 96819

ILLINOIS

Illinois Pest Control Association; Richard Jennings, Secretary; 18 South Michigan Avenue; Chicago, Illinois 60603

INDIANA

Indiana Pest Control Association; Dr. Gary W. Bennett, Secretary-Treasurer; c/o Entomology Hall, Room 112; Purdue University, Lafayette, Indiana 47907

IOWA

Iowa Pest Control Operators Association; Jim England, Secretary; c/o Presto-X-Company; 1225 South Saddle Creek Road; Omaha, Nebraska 68106

KANSAS

Kansas Termite Pest Control; Curtis C. Dozier, Secretary-Treasurer; c/o Pest Extermination Service; 910 West Douglas; Wichita, Kansas 67203

LOUISIANA

Louisiana Pest Control Association; Paul K. Adams, Secretary; c/o Adams Pest Control; Box 5006; Alexandria, Lousiana 71301

MAINE

New England Pest Control Association; Clarke C. Keenan, Secretary; 701 Washington Street; Braintree, Maine 02184

MARYLAND

Maryland Pest Control Association; William T. Fell, Secretary; c/o Fell's Pest Control Company; 3950 Falls Road; Baltimore, Maryland 21211

MICHIGAN

Michigan Pest Control Operators Association; Mrs. Jane Peden, Secretary-Treasurer; 408 W. Napier; Benton Harbor, Michigan 49022

MINNESOTA

Minnesota Pest Control Association; Michael W. Laughlin, Secretary; c/o Smith Laughlin Exterminating; 1910 University Avenue; St. Paul, Minnesota 55104.

MISSISSIPPI

Mississippi Pest Control Association; Mills L. Rogers, Secretary-Treasurer; c/o Rogers Entomological Service; P.O. Box 1084; Cleveland, Mississippi 38732

MISSOURI

Greater Kansas City Pest Control Association; Walter E. Mosier, Secretary-Treasurer; 712 West Blue Parkway; Lees Summit, Missouri 64063

Greater St. Louis Pest Control Association; Charles Zelsman, Secretary; c/o McCloud & Company; 127 B. Weldon Parkway; Maryland Heights, Missouri 63043

Missouri Pest Control Association; Charles R. Swender, Secretary-Treasurer; 3623 State Avenue; Kansas City, Kansas 66102

NEBRASKA

Nebraska State Pest Control Association; Monroe H. Usher, Secretary-Treasurer; 2416 N Street; Lincoln, Nebraska 68501

NEW JERSEY

New Jersey Pest Control Association; Fred Hubert, Executive; 8-15A Pinehurst Drive; Lakewood, New Jersey 08701

South Jersey Pest Control Association; Ms. Jannette Surace, Secretary; c/o La-Mar Exterminating Company; 643 South Orchard Road; Vineland, New Jersey 08360

NEW YORK

Long Island Pest Control Association; Howard P. Dewitt, Secretary-Treasurer; c/o Dewitt Pest Control; 36 Songsparrow Lane; Centereach, New York 11720

Professional Exterminators Association of Greater New York; Maury Eldridge, Secretary; 2070 Bronx Park East; Bronx, New York 10462

NORTH CAROLINA

North Carolina Pest Control Association; George Robbins, Secretary-Treasurer; c/o A-1 Termite and Pest; 919 Emerald Drive; Lenoir, North Carolina 28645

OHIO

Ohio Pest Control Association; John G. Breen, Secretary; c/o Buckeye Terminix; 2121 Riverside Drive; Columbus, Ohio 43221

OKLAHOMA

Oklahoma Pest Control Association; Mrs. Rosa A. Fisk, Secretary-Treasurer; 2600 SW 44th Street; Oklahoma City, Oklahoma 73119

OREGON

Oregon Pest Control Operators Association; P.O. Box 5004; Salem, Oregon 97304

PENNSYLVANIA

Pennsylvania Pest Control Association; William Kaighn, Executive Coordinator; 15 North York Road; Willow Grove, Pennsylvania 19090

SOUTH CAROLINA

South Carolina Pest Control Association; R. W. Garrison, Secretary-Treasurer; P.O. Box 443; Florence, South Carolina 29501

TENNESSEE

Tennessee Pest Control Association; Russell Bull, Secretary-Treasurer; c/o Russell's Exterminating; 4302 Papermill Road; Knoxville, Tennessee 37919

TEXAS

Greater Houston Pest Control Association; Tom DeLay, Secretary; c/o ALBO Exterminating Company; 3806 Arc; Houston, Texas 77063

Texas Pest Control Association; Don R. McCullough, Executive Director; 4302 Airport Boulevard; Austin, Texas 78722

VIRGINIA

Tidewater Pest Control Association; A. T. Gardner, Secretary-Treasurer; P.O. Box 7010; Norfolk, Virginia 23509

Virginia State Pest Control Association; Bruce B. Belfield, Secretary; c/o Universal Exterminating Company, Incorporated; Route 1, Box 603; Mechanicsville, Virginia 23111

WASHINGTON
(District of Columbia)
Washington Pest Control Association; Ralph Hughes, Secretary-Treasurer; c/o Paramount Pest Control; 110 Gordon Road; Falls Church, Virginia 22046

WASHINGTON
(State)
Washington State Pest Control Association; Christopher Senske, Secretary-Treasurer; c/o Senske Weed and Pest Control; P.O. Box 3024; Spokane, Washington 99220

WEST VIRGINIA
Pest Control Operators of West Virginia; Herman L. Hogan, Jr., Secretary-Treasurer; P.O. Box 9445; South Charleston, West Virginia 25309

WISCONSIN
Wisconsin Pest Control Association; Mrs. S. Ferrito, Secretary; c/o Safeway Pest Control; W140 59341 Boxhorn Drive; Hales Corner, Wisconsin 53130

ASSOCIATIONS AFFILIATED WITH NPCA

Services Offered Codes:	3. Termite Control	7. Weed Control
1. General Pest Control	5. Sale of Products	8. Bird Control
2. Fumigation	6. Ornamental	A. All Classifications

ABU DHABI
MYSODET PVT, LTD.; Sh Zayed 2nd Street; Box 2394; Abu Dhabi, UAE; K. C. Gopinath, Service Manager

ANTILLES
INTER-AMERICAN PEST CONTROL; Midden Straat, No. 4A; P.O. Box 779; Willemstad, Curacao; Ms. Lois Metsh; Affiliate of Truly Nolen

RAMCO PEST CONTROL; Sweelinck Straat 21; Oranjestad, Aruba; Netherlands Antilles; Roy A. Maduro, Owner-Manager (A)

ARGENTINA
CYGON SA; Aispurua 3150; Buenos Aires, Argentina; Mauricio Andre, President

AUSTRALIA
ABLE PEST CONTROL CO.; 12 Rosetta Street; West Croydon, Adelaide; South Australia; W. Monteath, Manager (1,2,3,7,8)

ADAMS PEST CONTROL PTY., LTD.; P.O. Box 205; Port Melbourne, Victoria; Australia 3207; Alan N. Banks, General Manager (A)

Other same name locations:

P.O. Box 34; Kilkenny, 5009; South Australia; J. S. McCormack

ADELAIDE PEST CONTROL PTY., LTD.; P.O. Box 60; Walkerville, SA, 5081; Australia; H. D. Bonney, Managing Director

ALLPEST—AUST—PTY., LTD.; 66B Canning Highway; Victoria Park, Perth, 6100; Australia; G. R. Peek, Managing Director (A)

Other same name locations:

196 Anzac Highway; Plympton, 538; South Australia K. W. Cannon, Branch Manager (A)

218 Old Cleveland Road; Cooparoo, Brisbane, 4151; (Queensland) Australia S. Adams, Branch Manager (A)

AMALGAMATED PEST CONTROL; 200 Beatty Road; Coopers Plains, 4108; Australia; John McCarron

BONNEY'S PEST CONTROL PTY, LTD.; 77 Lime Avenue, Box 759; Mildura Vic., 3500; Australia W. D. Medhurst, Managing Director (A) B. A. Scott, Managing Director

172

DAWSON'S PESTAF*/AND WEED CONTROL P/C; 36 Clive
Street, West Footscray; Victoria, Australia 3012
Charles M. Dawson, Owner-Operator   (1,3,5,7)

W. A. FLICK & CO. PTY., LTD.; 73 Victoria Avenue;
Chatswood NSW, 2067; Australia
D. A. Flick, Director   (A)

R. J. GRINHAM LIMITED; 13 Doris Street, Picnic Point;
Sydney NSN, Australia
Ron J. Grinham, Director   (1,2,3,5,6,7,8)

H.P.C. PTY., LTD.; 47 Burswood Road, Victoria Park; West
Australia, 6100
Frank L. Bonney   (1,2,3,5,6,7)

MAXWELL ROBINSON & PHELPS; 105 Cambridge Street;
West Leederville; West Australia
R. S. Phelps, Manager   (1,3,6,7,8)

RENTOKIL PTY., LIMITED; P.O. Box 72; Artarmon NSW,
2064; Australia
M. J. Clarke, Managing Director   (A)
P. R. Meadows, Technical Director

BAHAMAS
TROPICAL EXTERMINATORS, LTD.; P.O. Box N-1388;
Nassau, Bahamas; (809) 322-2157
G. Thomas Sweeting, General Manager (1,2,3,5,6,7)

BAHRAIN
JISHI PEST CONTROL; Al-Khalifa Road; P.O. Box 617;
Bahrain, Arabian Gulf; A. S. Nair

BANGLADESH
HERBERTSONS BANGLADESH, LTD.; 14 Sadarghat Road;
P.O. Box 112; Chittagong, Bangladesh; Firoze Kabeer

BERMUDA
BERMUDA PEST CONTROL; P.O. Box 287, Paget 6; Bermuda;
(809) 295-1823
Denis De Sousa, Owner   (1,2,3,5,8)
Gary Soares, Manager

173

MID OCEAN PEST CONTROL; P.O. Box 105; Paget 6;
Bermuda; (809) 295-5730
Edward D. Finnerty, Owner-Manager   (1,2,3,8)

BRAZIL

ABC EXPURGO, LTDA.; P.O. Box 767; São Bernardo do
Campo; São Paulo, Brazil; Ms. Lucia Schuller, Exec. Vice
President; Peter Schuller, President

ELECTROLUX SERVICOS SOCIEDADE COMMERCIAL,
LTDA.; Av. São Gabriel, No. 55, 7th Floor; São Paulo, Brazil
Aldo De Avila Jr., Manager

CANADA

ABELL WACO LIMITED; 426 Attwell Drive; Rexdale, Ontario;
Canada M9W 5B4; (416) 675-1635   (A)
Ralph E. Abell, President; Jim McConnell, Vice President; John
R. Abell, Comptroller

AIR GUARD PEST CONTROL SERVICE; 2345 Hensall Street;
Mississauga, Ontario; Canada L5A 2T1;
(416) 661-1313   (A)
J. F. Chandler, President

Other same name locations:

38 Avon Road; Kitchener, Ontario; Canada N2B 1T6;
(519) 745-4037   (A)
G. C. Chandler, Vice President; R. Chandler, Manager

1941 Edenvale Cres; Burlington, Ontario; Canada L7P 3HP;
(416) 637-8767   (A)
J. H. Solomon, Vice President

B.B. EXTERMINATING, INC.: C P 1034; Shawinigan, South
Quebec; Canada G9P 4E6
(819) 536-3808   (1,5,8)
R. E. Bellemare

B.C. PEST CONTROL, LTD.; 2511 West Broadway; Vancouver,
B.C.; Canada V6K 2E9
(604) 731-2711   (1,3,8)
J. Van, President; R. Kind, Secretary

BURTON'S SANITATION, LTD.; P.O. Box 421; Kingston,
Ontario; Canada;
(613) 546-6641   (A)
C. S. Burton, President; T. Burton, General Manager

CAMERON EXTERMINATION; 918 McManamy Boulevard;
Sherbrooke, P.Q., Canada J1H 2N3;
(819) 569-2847   (1)
Gerard Cameron, Owner

EXTERMINATION TRANS METROPOLE PEST CONTROL
LTEE., LTD.; 360 Michel Jasmin, Dorval; Quebec, Canada H9P
1C1;
(514) 631-1772   (1,6,8)
J. P. Lamy, President

GENERAL PEST CONTROL LIMITED; R.R. #6, 42 Easton
Road; Branford, Ontario; Canada N3P IJ5;
(519) 756-6641   (1,2,5,8)
Gilbert Gilroy, General Manager

KEMSAN, INC.; P.O. Box 727; Oakville, Ontario; Canada L6J
5C1;
(416) 845-2271   (5)
Robert D. Paul, General Manager

LEO SEVIGNY & FILS, INC.; 2949 Chemin Ste.-Foy; Quebec,
Canada;
(418) 653-3101; Serge Sevigny

MAHEU & MAHEU, INC.; 319 Du Pont Street; Quebec, P.Q.;
Canada G1K 6M2;
(418) 525-4755   (1,2,5,6,7,8)
Paul Maheu, President

MAHEU EXTERMINATION, LTD.; 3128 Masson; Montreal,
P.Q.; Canada H1Y 1X8
(514) 725-6489   (1,5,8)
Michael Bissonnette, Vice President

MYSTO INCORPORATED; 3333 Cremazie Boulevard East;
Montreal, Quebec; Canada H1Z 2H8;
(514) 721-4921   (1,2,5,8)
Pierre F. Aubry, Exec. Vice President

NFLD SANITARY SERVICE AND EXTERMINATING; P.O.
Box 55, Stephenville; Newfoundland, Canada; (709) 643-2014;
George Banfield

OLYMPIC PEST CONTROL, LTD.; 1204 10th Avenue; New
Westminster, B.C.; Canada V3M 3H8;
(604) 525-3633   (1,2,3)
Rex Case, President; Bob Brady, Supervisor; K. Ferguson,
Supervisor

P E I PEST CONTROL LIMITED; 160 Brackley Point Road;
Sherwood-Prince Edward Island; Canada C1A 6Y9;
(902) 894-9740   (1)
Frank Legault, President

PCO SERVICES LIMITED; 232 Norseman Street; Toronto,
Ontario; Canada M8Z 2R4;
(416) 231-7277   (1,2,3,5,8)
Alf H. Gartner, President; K. Spencer, Vice President

Other same name locations:

1472 Kingston Road; Toronto, Ontario; Canada M1N 1R6
(416) 699-9373   (1,2,3,5,8)
B. Carleton, Service Manager

6791 East Hastings; Burnaby, B.C.; Canada V5T 1S6
(604) 291-7741   (1,2,3,5,8)
D. C. Smith, General Manager

10044–158th Street; Edmonton, Alberta; Canada T5P 1A4
(403) 483-3070   (1,2,3,5,8)
T. Bisanz, District Manager

196 Parkdale Ave. North; Hamilton, Ontario; Canada L8L 2X5
(416) 544-6242   (1,2,3,5,8)
S. McKenzie, Service Manager

71 Portland Street; Toronto, Ontario; Canada M8Y 1A6
(416) 252-7261   (1,2,3,5,8)
Ms. Rose Skala, Service Manager

1506 Fairburn Avenue, Sudbury, Ontario; Canada P3A IN7
(705) 566-8100
P. Cloutier, Service Manager

489 Highbury Avenue; London, Ontario; Canada N5W 4K2
(519) 452-3730   (1,2,3,5,8)
Harvey Hackland, Service Manager

71 Portland Street; Toronto, Ontario; Canada M8Y 1A6
(416) 252-6411   (1,2,3,5,8)
J. Thurling, Service Manager; C. Acheson, Regional Manager

273 Spadina Avenue; Toronto, Ontario; Canada M5T 2E3; (416)
862-8333; H. Ho Ken, Service Manager

461 Dunlop Street, West; Barrie, Ontario; Canada; (705) 728-7101;
G. Muldoon, Service Manager

3550 Wolfedale Road; Mississauga, Ontario; Canada L5C 2V6;
(416) 276-5404; F. Griffin, Service Manager; J. Farrell, Regional
Manager

PCO SERVICES QUE., LIMITED; 6490 Bombardier Street;
Montreal, Quebec; Canada H1P 1E2; (514) 326-6300
P. Dubreuil, President   (A)

THE PESTCO COMPANY OF CANADA LIMITED; 134
Doncaster Avenue, Thornhill; Ontario, Canada L3T 1L3; (416)
881-5900
E. Valder, President   (1,2,3)

PESTROY CHEMICAL COMPANY, LTD.; 1655 Edouard
Laurin Boulevard; St. Laurent, Montreal, PQ.; Canada H4L 2B6;
(514) 336-6110
Norman V. J. Smith, Secretary-Treasurer   (A)

PHOSTOXIN SALES OF CANADA, LTD.; P.O. Box 343;
Winnipeg, Manitoba; Canada R3C 2H6; (204) 943-7796; Dr. Ernest
A. Liscombe, President; David R. Liscombe, Secretary

PIED PIPER COMPANY LIMITED; 7061 Gilley Avenue;
Burnaby, B.C.; Canada V5J 4X1; (604) 434-6641
George R. Turner, President   (A)

PMS PEST MANAGEMENT SERVICE, LTD.; 934 Brunette
Avenue; Coquitlam, B.C.; Canada V3; (604) 524-5511
Val J. Helgason, President   (1,2,3,5,8)

POULIN PEST CONTROL; 24 Marion Place; Winnipeg,
Manitoba; Canada R2H 0S9
Donald Poulin, President   (A)
Claude Clement, General Manager

RELIABLE EXTERMINATORS; 1730 McPherson Court;
Pickering, Ontario; Canada L1W 3E6

SAFEWAY PEST CONTROL SERVICE; 300 Sydney Street S;
Kitchener, Ontario; Canada N2G 3W1; (519) 576-6592
Harry J. Hutt, Owner   (1,3)

SANEX CHEMICALS, LTD.; 6439 Netherhart Road;
Mississauga, Ontario; Canada L5T 1C3; (416) 677-4890
William Brennan, President   (5)
R. L. Prindiville, Sales Manager

Other same name locations:

6490 Bombardier Street; Montreal, Quebec; Canada H1P 1E2;
(514) 323-8221; Robert Desjardins, Sales Manager

SANIBEC CORPORATION; 325 Belvedere South; Sherbrooke,
Quebec; Canada J1H 4B6; (819) 565-8688
Leonard T. Laflamme, Manager   (1)

VANCOUVER FUMIGATING CO., LTD.; P.O. Box 2009;
Vancouver, B.C.; Canada V6B 3R2; (604) 273-6321
W. Moscaliuk, Manager   (1,2)

WESTERN EXTERMINATING CO., LTD.; 9795 Verville
Street; Montreal, P.Q.,; Canada H3L 3E1; (514) 384-6550
Edward Rudick, General Manager (1,2,5,8)

WIPP PEST CONTROL COMPANY; 468 Pitt Street East;
Windsor, Ontario; Canada N9A 2V8
Arthur L. Wippman   (A)

CEYLON
CEYLON PEST CONTROL COMPANY; 34 ½ Bailie Street;
Colombo 1, Ceylon
Desmond Gunawardena   (1,2,3,6,7)

CHILE
AGRICOLA FUDESAM, LTDA.; Ahumada 312, Ofc. 711;
Santiago, Chile
Patricio M. Dufour, Agricultural Engineer   (1,2,3,5)
Oscar G. Glasinovic, Agricultural Engineer

COLOMBIA
FUMIGAX DE COLOMBIA, LTDA.; Calle 33 56-70, Barrio Gran
Av.; P.O. Box 3069, Medellin; Colombia, South America
Rodrigo P. Posada   (1,2,3,5,6,8)

DENMARK
A-S RENTOKIL; Bogholder Alle 40; 2720 Vanlose; Copenhagen,
Denmark
F. Andersen, Manager   (A)

MORTALIN; DK-4690, Haslev; Denmark
Aksel Hoff, President   (A)

SKADEDYRCENTRALEN; Hogevej 11; DK 6700, Esbjerg;
Denmark
A. L. Lange   (1,2,3,5)

DOMINICAN REPUBLIC
LA EXTERMINADORA, C Por A; Avenida de los Proceres 3;
P.O. Box 1193; Santo Domingo; Dominican Republic
Jose Batlle Nicolas, President   (1)

FUMIGADORA ORIENTAL, S.A.; Duarte #2, P.O. Box 348; La
Romana, Dominican Republic; Francisco Torres, President

TRULY NOLEN DOMINICANA; Apartado Postal 36-2; Santo
Domingo; Dominican Republic; Jose A. Rey
(Other location of TRULY NOLEN)

ECUADOR
FUMIGADAZA, S.A.; P.O. Box 4653; Guayaquil, Ecuador;
Victor Pino Y

EL SALVADOR
COMPANIA EXTERMINADORA DE INSECTOS Y
ROEDORES; Calle Gerardo Barrios #1722; San Salvador
Salvador Parras, President   (1,2,3,6,7,8)

ENGLAND
See UNITED KINGDOM

FRANCE

AMBOILE CHIMIE; 94490 Ormesson-Sur-Marne; France; (A)
A. J. Santarelli, Laboratory Manager

LABORATOIRE AGUETTANT; 1 Avenue Jules-Carteret; 69
Lyon 7; France; Georges Aguettant

SOCIETE L'ETOILE SARL; 201 Rue Du Faubourg Sainte-
Honore; Paris 8me,
France   (1,2,3,5,7,8)
N. A. Shimshi, Managing Director

GERMANY

THE FARM RESEARCH CORPORATION; Obere Flueh 6;
Saeckingen; West Germany 7880   (1,7)
Dr. Peter G. Wilde, President

HEERDT-LINGLER GMBH; Deutsche Gesellschaft fur;
Schadlingsbekampfung MBH, 6000 Frankfurt 1; West Germany;
Klaus Baumert

PEST CONTROL GMBH; P.O. Box 2067; 6380 Bad Homburg 1;
West Germany   (1)
Lothar Kickert; Dr. G. Voelz

PREVENTA, GMBH; North-Western Region; Siemensstr 4, 4040
Neuss 21; West Germany   (1,5,7,8)
Peter H. Uhlig, General Manager

Other same name locations:

Southern Region; Kernerstr 22A, 7000 Stuttgart;
West Germany (1,5,7,8)
Dieter Zingler, General Manager

RENTOKIL GMBH; Oberratherstr 10; 4000 Dusseldorf 30; West
Germany; Heine Jonker, General Manager

Other same name locations:

Assar-gabrielsson-Str 20; 6057 Dietzenback-Steinway; Germany;
Bob Dow

808 APPARATE UND PRAPARATE; W. Frowein; D-7470
Albstadt-Ebingen; West Germany; H. J. Frowein (5)

GUATEMALA

EXTERMINADORA SELLES DE GUATEMALA, S.A.; Via 7,
5-33, Zona 4; Guatemala City,
Guatemala (1,2,3,6,7,8)
Carlos E. Selles, General Manager

HOLLAND

B.V. RENTOKIL CHEMIE; Volmerlaan 9; 2288 GD-Rijswijk;
The Netherlands (A)
J. D. Uding, General Manager

HONG KONG

PAUL INTERNATIONAL; P.O. Box 95357; 16 Humphrey's
Avenue, 4/F,
Kowloon; Hong Kong (1,2,3,5)
Paul T. Owen, President

INDIA

ANGLO ORIENT INSECTICIDE SVCS; Bhupen Chambers,
2nd Floor; Dalal Street; Bombay 400023, India (1,2,3,6,7,8)
Gajendra Bhagat

BANGALORE PEST CONTROL CORP.; 4/164 V Main Road, Chamrajpet; PB 1814, Bangalore 560018; India; R. Janardanan, General Manager

CHEMAFUMES PRIVATE LIMITED; Dena Bank Building; 17, Horniman Circle; Bombay 400 023, India   (1,2,3,5) Manik D. Lotlikar, Managing Director

FILCCO PRIVATE, LTD.; Sewree Fort Road, Behind Minerva; Studios Sewree—East; Bombay 15, India   (1,2) V. H. Pancholi

HARDY'S PEST CONTROL SERVICING REGISTERED; 1620, Bahadurgarh Road; Delhi 110006, India   (1,2,3) Y. S. Behl, Proprietor

Other same name locations:

65-A, Gandhi Nagar; Jammu, India; V. K. Chopra, Branch Manager

JARDINE HENDERSON, LTD.; 187 Lloyds Road; Madras 600 086, India; I. Raman, Manager

B. KAIKHUSHROO & COMPANY; 19, Bank Street, 2nd Floor, Fort, Bombay 400 001; India   (1,3,6) D. P. Siganporia

PEST CONTROL INDIA PRIVATE, LTD.; 36 Yusuf Bldg., Flora Fountain; Mathatma Gandhi Road, Fort; Bombay 400 023, India   (A) N. S. Rao, Chairman

Other same name locations:

7, Jantar Mantar Road; New Delhi 110001, India; Mohan R. Bajikar

PEST CONTROL, M. WALSHE; Oriental House; 7 J Tata Road; Bombay 400 020, India   (A) P. S. Pruthi

PESTOGONE; 8/47 Ellora Park, Race Course; Circle, Baroda
390007; Gujarat, India   (1,2,3,6,7,8)
Anant V. Nene, Partner

INDONESIA

C. V. GADING; Jalan Setia # 15, Jakarta-Timur; Jakarta,
Indonesia   (1,2,3,6,7)
Henky S. Pryana, Director

MENDALA PEST CONTROL; Jalan Jaya Mendala II/2;
Pertamina Oil Village, Pancoran; Jakarta, Selatan, Indonesia;
Joe Hok Tan, Owner-Operator   (1,3)

PISOK PEST CONTROL & SANITATION; Taman Kimia 8;
P.O. Box 2332; Jakarta, Indonesia; M. J. Tampenawas, Managing
Director; M. Tanjung, Entomologist

ISRAEL

A. MOSKOVITS & SONS, LTD.; P.O. Box 8032; Haifa 31080,
Israel   (A)
Arie Moskovits, Director-Owner; Yigal Merav, Owner-Operator

MAKHTESHIM—AGAN; Export Office; P.O. Box 60; Beersheva,
Israel; J. Y. Rein, Export Manager

NATIONWIDE EXTG., ISRAEL; Modiin Street 18—POB 2272;
Ramat Gan, Israel   (A)
E. Amichai, Owner-Operator

JAPAN

APEX SANGYO COMPANY, LTD.; 12-16, Shiba-Koen, 2-
Chome; Minato-Ku, Tokyo; Japan   (A)
Mikio Motoki, Chairman; Mitsugu Motoki, President

DUSKIN COMPANY, LIMITED; Sekaicho Building, 6-24;
Nakatsu 1-Chome, Oyodo-Ku; Osaka 531, Japan; Seiichi Suzuki,
President

HIROSHIMA-KEN YAKUGYO CO., LTD.; 11-8 Shoko-Center,
3-Chome; Hiroshima City 733; Japan   (1,3,5,8)
Masayoshi Yoshimura, President

HOHTO SHOJI CO., LTD.; Aida Bldg., No. 20; 1-Chome,
Yotsuya Shinjuku-Ku; Tokyo, Japan   (5)
H. Shiboh, President

THE IKARI LTD.; No. 23-7 Shinjuku, 3-Chome; Shinjuku-Ku
Tokyo, Japan   (1,2,3,5,7,8)
Toshishige Kurosawa, President

TAIYO KAKO COMPANY, LIMITED; 520 Kamishima, Ohito-
Cho; Tagata-Gun, Shizuoka Pref.
410-23 Japan
Takeshi Nishijima, President; Mizuki Nishijima   (1,3,6,7,8)

TEISO KASEI COMPANY; 164-1 Nishizima; Shizuoka 422; Japan
Taneo Yoshida, President   (1,2,3,6,8)

TERMITE INSURANCE ECONOMY ORG.; 2-2-10
Uenosakuragi; Taitoku, Tokyo; Japan 110; Kiyoshi Yanagisawa,
President

YUKO CHEMICAL INDUSTRIES CO.; 2-123 Tsutoiidencho;
Nishinomiya, Hyogo; Japan   (1,5)
Minoru Maeda, President

KOREA

CHYUNU PEST CONTROL CO., LTD.; 243 Block 1 Sinsa-
Dong; Gangnam-Gu, Seoul, 134-03; Republic of Korea; Soon P.
Chyun, President

KOREAN PCO COMPANY, LTD.; 82-14 4Ga Jung ang-Dong;
Jung-Gu Busan; Korea   (1,2,3,5,8)
Seong H. Park, President

MALAYSIA

THOMAS COWAN & CO. PTE., LTD.; 15A Lorong Universiti
'C'; Petaling Jaya, Selangor; Malaysia   (1,3)
Jimmy Hoe, Manager

MEXICO

FUMIGACIONES SANMEX; Garibaldi 1059; Guadalajara,
Jalisco; Mexico; L. Mexin

RODEX, S.A.; Reforma 308–4th Floor; Mexico 6; Mexico; (905) 533-6423 (A)

Jorge Martinez, President

Other same name locations:

Horacio 545, Mexico 5; Mexico; (905) 545-6755; Enrique Macias

TRULY NOLEN; Box 564, Hermosillo Sonora; Mexico; Juan Gonzales
(Other location of TRULY NOLEN)

NEW ZEALAND

RENTOKIL LIMITED; P.O. Box 13-445, Onehunga; Auckland 6, New Zealand (A)

B. H. Phipps, Managing Director

NIGERIA

PESTKIL (NIGERIA) LIMITED; P.O. Box 8190; Lagos, Nigeria; W. Mennel, Managing Director

OMAN

PESTOP FUMIGATION SERVICE; P.O. Box 937; Muscat Sultanate of Oman (1,2,3)

A. M. Shikely, Manager

PAKISTAN

CENTRAL IMPERIAL CHEMICAL, IND.; 19–Sunny Plaza; Hasrat Mohani Road; Karachi, Pakistan (2,3)

Abdul Samad Khan

D. M. PEST CONTROL; 128/3 Off Maneckji Street; Garden East, Karachi-3; Pakistan (A)

Kassemali D. Jafferbhoy, Partner

PANAMA

TRULY NOLEN DE PANAMA; Apartado 7074, Avenida Mejieo; Calle 31, Panama 5; Republica de Panama; Virgilio Gomez
(Other location of TRULY NOLEN)

PHILIPPINES

EMAR MERCHANDISING & SERVICES; 151 Chico Street
Project 2; Quezon City; Philippines; Marcelo Roa Yu, Owner

SAUDI ARABIA

AL HOTY PEST CONTROL; P.O. Box 446; Dhahran Airport;
Saudi Arabia; Adel Abdalla Abdalla, Manager

AL-SAHIB GENERAL SERVICE ESTABLISHMENT; P.O.
Box 201, Dhahran Airport;
Saudi Arabia   (1,2,3,5,6,7)
Fouad M. Saleh, Manager

ARABIAN AMERICAN OIL COMPANY; Medical Dept., Box 76
(K103); Dhahran, Saudi Arabia; Lee M. Carson, Supervisor

SINGAPORE

THOMAS COWAN & CO. PTE., LTD.; First Floor, Inchcape
House; 450-452, Alexandra Road; Singapore, 5   (1,3)
John Ewing, Director; Robert Wiener, Director; Robert Lee,
Office Manager; Teoh Kian Seng, Manager; Andrew Tan,
Operations Manager

THE WELLCOME FOUNDATION, LTD.; South East Asia
Area, Jurong; P.O. Box 2; Singapore 22;   (5)
J. M. Sanders, Marketing Manager

SPAIN

CORPORACION INTERNACIONAL RATIN S.A.; Virgen De
Gracia 7; Barcelona 6, Spain   (A)
Francisco Llagostera, Director

LABORATORIOS SAMDI S.A.; Menendez Pelayo 53, 2E;
Madrid 9, Spain; Jose S. Carrion, President; Jose Marcial, Jr.,
Director

S E D Y F; Generalisimo 16-Bajos; Tarragona, Spain   (A)
Jose Sendra, Director

SERVI PLAGA, S.A.; Infanta Mercedes, 92; Oficina 1.12; Madrid 20, Spain; Pedro Rodriguez, President

SWEDEN

ANTICIMEX AB; Vasagatan 46; Box 726; 10130 Stockholm, Sweden (A)
Uno Ljungberg

TANZANIA

TANZANIA PEST CONTROL COMPANY; P.O. Box 20990; Dar es Salaam, Tanzania; K. P. Rawal, Owner

THAILAND

KERR INTERNATIONAL SERVICES COMPANY, LTD.; 37 Soi 12 Sukhumvit Road; Bangkok, Thailand (1,2,3,4)
Maurice H. Kerr, Managing Director

KING SERVICE CENTER CO., LTD.; 229 Sukhumvit 55 Thong Lor 11; Bangkok 11, Thailand (1,2,3,5)
Suchart Leelayouthyotin, Managing Director

PAE THAILAND CO., LTD.; 20/6 Soi Shidlom; Bangkok Bazaar; Bangkok, Thailand; Samroeng Puagpipat, Manager

UNITED PEST CONTROL CO., LTD.; Larn Luang Road; 29277 Sapan Kao Trade Centre; Bangkok 3, Thailand (A)
Charles S. Phenpimon, General Manager

WELLCOME THAILAND, LTD.; Wellcome Pest Control Divison; P.O. Box 2475; Bangkok, Thailand (1,2,3,6)
R. D. Shaw, Managing Director; Visooth Lohitnavy, Manager

UNITED KINGDOM

RENTOKIL, LTD.; Felcourt, East Grinstead; Sussex, RH19 2JY,
United Kingdom (1,2,3,5,7,8)
A. A. Tyrer, Managing Director

VERMINEX, LTD.; 11 Guilford Street; London WC1N 1DU;
England   (1,2,5,8)
S. P. Egleton, Director

WELLCOME FOUNDATION, LTD.; 183 Euston Road; London,
England NW1 2BP   (5)
Martin Evans, Planning Manager

VENEZUELA

OFICINA TECNICA MOORE C.A.; Calle San Camilo, No.
03-34; Av. Las Palmas, Los Caobos; Caracas 105, Venezuela (A)
Nilo A. Castro, President

Other same name locations:

Calle 107 (Navas Spinola); No 98-44; Valencia, Venezuela;
Alejandro Martinez, General Manager

TERMINEX C. A.; Apartado 1525; Caracas, Venezuela (A)
Jose Calvino, President

WEST INDIES

CARIBBEAN PEST CONTROL, LTD.; Carmichael House; St.
George, Barbados; West Indies   (A)
Henry Vieira, Managing Director

TRINIDAD PEST CONTROL, LTD.; 91 Cascade Road; Cascade,
Trinidad; West Indies   (A)
R. D. Jeary, Managing Director

BAYER CARIBBEAN, LTD.; P.O. Box 679C; Lowland Christ
Church; Barbados, West Indies; H. E. Bernhardt, Managing
Director

INDEX

f after a page number indicates reference is to a figure.

190

roaches of, 59
Electrical appliances, damaged by
   cockroaches, 59, 62
English roaches, environment for,
   41f
Environmental conditions, cockroaches and, 16, 41, 43f, 79f, 80f,
   81f; grouping and, 44
Environmental Protection Agency,
   standards for insecticides, 110, 161
Eyelashes and hair, eaten by cockroaches, 17, 19, 39, 47

Fecal matter, 41; cockroach species
   identification and, 45
Fingernails and toenails, eaten by
   cockroaches, 17, 19, 39, 47
Fire hazards, aerosols as, 111, 150
Fishtanks, during de-roaching, 111,
   162
Flying roaches, 41, 64f
Fogging and misting, 108, 157
Food, of cockroaches, 16, 39, 40f, 73,
   74f. *See also* Food supplies
Food-processing facilities, de-roach-
   ing instructions for, 121–122
Food supplies, cockroach con-
   tamination of, 19, 21, 22f, 23, 163
Fossil cockroaches, 50; compared
   with present-day species, 71; loca-
   tions in North America, 16
French cockroach, 30f

Gecko lizards, cockroaches and, 58
German cockroach, 29f, 48f, 89f, 90f;
   breeding rate of, 62, 155; dwelling
   places of, 41; entry into houses by,
   99; evolution and spread of, 16;
   facts on, 85; longevity, 47
Gromphadorhina spp, 55

Health: cockroaches as danger to,
   23; and pesticide inhalation, 161
Hissing cockroaches, 55
Hospitals, de-roaching methods for,
   124
Housekeeping practices, 44, 49

Insect phobia, 46–47

Italian cockroach, 32f

Japanese cockroach, 34f

Kepone, 155
Killmaster, 112
Kitchens, de-roaching instructions
   for, 107, 118–119
Knockdown agents, 149, 150

Legs of cockroaches, 71; loss of, 73
Licensing and certification of pest-
   control operators, 47
Liquid sprays and emulsions, 152f,
   153
Living rooms, de-roaching of,
   119–120

Maze travel by cockroaches, 45;
   temperature and, 59
Medicine, use of cockroaches in, 17,
   18f, 19
Mexican cockroach, 35f
Misting. *See* Fogging and misting
Molting by cockroaches, 42, 76
Mutations, 16

Negative phototropism, 41
Nymphs, 42, 57, 76, 78f

Odor of cockroaches, 21, 23
Office buildings, de-roaching of,
   120–121
Oriental cockroach, 91f, 92f; habits
   of, 41; longevity of, 47; mislabeled
   as "water bugs," 39, 41; profile of,
   87
Ovipositor, evolution of, 16

Paleoblattidae, 11
Palmetto bug, 41
Parthenogenetic reproduction, 73
Periplaneta americana. See
   American cockroach
Pest-control operators, 165f; charges
   of, 106; and control vs. eradica-
   tion, 26; guarantees of, 112;
   licensing and certification of, 47;
   and roach eradication, 44; selec-

Pest-Control Operators (*cont.*)
tion of, 105; service contracts
with, 106–107. *See also* De-roaching; Pesticides
Pesticides: accidental contamination
with, 162; applied with paint
brushes, 112; avoidance by
roaches, 154f, 158f, 159f; choice of,
97, 110–111; common, chemical,
and trade names of, 151; concentrations of active ingredients
in, 110; effects on pets, 162; inhalation of, 161; odors of, 161–162;
resistance to, 16, 147f; safety of
children and pets and, 161. *See
also* Aerosols; Baits; Bombs; Fogging and misting; Liquid sprays or
emulsions; Roach tape; Traps;
ULV
Pet shops and zoos, de-roaching
methods for, 123
Pets: aerosol bombs and, 111; effects
of pesticides on, 162; roach traps
and, 111; and safety tips for
pesticide use, 161
Pheromones of cockroaches, 59
Professional pest-control associations, 105; international affiliates
of National Pest Control Association, 171–188; listed by states,
167–171
Pyrethrin, 108, 149, 150, 153; effects
on cockroach behavior, 162–163;
inhalation of, 161

Radiation, cockroaches and, 16, 49
Radios, cockroach infestation of, 59,
62
Rain, cockroaches predicting, 55, 57
Residual pesticides, 97, 108,
149–150, 163
Restaurants: roach control in, 107;
roaches in, 49. *See also* Food-
processing facilities
Roach Motel, 111, 155
"Roach" smoking, 67f
Roach tape, 157
Roach traps, 111, 117, 155

"Roaches Are Here to Stay" exhibit,
59
Russian cockroach, 33f

Schools, de-roaching of, 125
Service contracts, 106–107
Sewers, cockroaches in, 55, 57
Ships, cockroaches on, 16, 17, 19, 26
Siamese cockroaches, 36f
Silica gel, 153
Soap, cockroaches eating, 45
Spanish cockroach, 28f
Sprays. *See* Aerosols; Liquid sprays
and emulsions
Steel wool, around pipes, 44, 46
Supella longipalpa. See Brown-
banded cockroach
Supermarkets. *See* Food-processing
facilities
Swimming, by cockroaches, 45, 55,
56f

Toothbrushes, contamination of, 162
Toothpaste, eaten by cockroaches,
45
Traps, 155
Tropical fish and birds, pesticides
and, 162

ULD. *See* ULV
Ultra Low Dosage. *See* ULV
Ultra Low Volume. *See* ULV
ULV, 157

VAPONA, 150
Vehicles, de-roaching, 62, 124

"Water bugs," 39, 41, 95
Water pipes, roaches transported
through, 99, 102f
Whistling or hissing cockroaches, 55,
61f
Wings of cockroaches, 11
Women: as professional exterminators, 59; researchers on cockroaches, 58–59

Zoos. *See* Pet shops and zoos